The Story of Manned Space Stations
An Introduction

D0890321

Philip Baker

The Story of Manned Space Stations

An Introduction

Published in association with
Praxis Publishing
Chichester, UK

Philip Baker
Church Crookham
Hampshire
UK

SPRINGER–PRAXIS BOOKS IN SPACE EXPLORATION
SUBJECT *ADVISORY EDITOR*: John Mason, M.Sc., B.Sc., Ph.D.

ISBN 978-0-387-30775-6 Springer Berlin Heidelberg New York

Springer is part of Springer-Science + Business Media (springer.com)

Library of Congress Control Number: 2007922811

Apart from any fair dealing for the purposes of research or private study, or criticism
or review, as permitted under the Copyright, Designs and Patents Act 1988, this
publication may only be reproduced, stored or transmitted, in any form or by any
means, with the prior permission in writing of the publishers, or in the case of
reprographic reproduction in accordance with the terms of licences issued by the
Copyright Licensing Agency. Enquiries concerning reproduction outside those terms
should be sent to the publishers.

© Praxis Publishing Ltd, Chichester, UK, 2007
Printed in Germany

The use of general descriptive names, registered names, trademarks, etc. in this
publication does not imply, even in the absence of a specific statement, that such
names are exempt from the relevant protective laws and regulations and therefore free
for general use.

Cover design: Jim Wilkie
Project management: Originator Publishing Services Ltd, Gt Yarmouth, Norfolk, UK

Printed on acid-free paper

Contents

vi   **Contents**

For Helen
Thank you, I owe you more than I can ever repay . . .

For Mum and Dad
Thank you for your support, always

To the staff of King's College Hospital Liver Unit
Without you this book would not have been written

Acknowledgements

As is usual with such a project, I have many people to thank.

First, Colin Burgess who I have to thank, or blame, for getting me started with this book. Your support and encouragement have always been very important.

I also need to thank my fellow Praxis authors, David Shayler and David Harland for their support, as well as the understanding of my publisher Clive Horwood who has led this first-time author through the minefield of book publishing.

Some of the images in this book have come from a free space flight simulator called "Orbiter". This software has allowed me to include images of spacecraft for which no images exist, or at least are of poor quality. Please visit *www.orbitersim.com* for more information. I would particularly like to thank David Polan, who has provided several such images for this book, including one on the front cover.

I mentioned in my dedication the staff of King's College Hospital Liver Unit, and I would now like to add to that by mentioning Dr. Micheal Hinnehan and his team, and Dr. John Ramage and his team at North Hampshire Hospital, both of whom helped me through the most difficult period of my life. As an aside, I would urge all U.K. based readers to register as organ donors at *www.uktransplant.org.uk*, because quite literally you would not be reading this book if someone had not done the same for me.

Figures

Color plates

The color plates can be found between pages 78 and 79

Abbreviations and acronyms

AAP	Apollo Applications Program
ARPS	Aerospace Research Pilot School
ASAT	anti-satellite
ASTP	Apollo–Soyuz Test Project
ATM	Apollo Telescope Mount
ATV	Automated Transfer Vehicle
BIS	British Interplanetary Society
CEV	Crew Exploration Vehicle
CMG	Control Moment Gyroscope
CRV	Crew Return Vehicle
CSM	Command & Service Module
DOR	Director of Operations
EOR	Earth Orbit Rendezvous
EDS	Earth Departure Stage
ESA	European Space Agency
ET	External Tank
EVA	Extravehicular Activity
EDO	Extended Duration Orbiter
EKG	Electrocardiogram
FGB	Functional Cargo Block (Russian acronym)
ISS	International Space Station
KSC	Kennedy Space Center
LM	Lunar Module
LOR	Lunar Orbit Rendezvous
LSAM	Lunar Surface Access Module
LV	Launch Vehicle
MODS	Manned Orbital Development Station
MOL	Manned Orbiting Laboratory

xvi **Abbreviations and acronyms**

MIT	Massachusetts Institute of Technology
MKBS	Early Soviet space station design
OS	Orbital Station
OSP	Orbital Space Plane
OWS	Orbital WorkShop
RCS	Reaction Control System
SMEAT	Skylab Medical Experiments Altitude Test
SFOG	Solid Fuel Oxygen Generators
SRB	Solid Rocket Booster
SSESM	Spent Stage Experiment Support Modules
SPS	Service Propulsion System
TKS	Heavy Space Station (Transport and Supply Craft)
TORU	remote-control flight system (russian acronym)
TOSZ	Heavy Orbital Station
VAB	Vehicle Assembly Building

Introduction

In 1971, Viktor Patsayez gazed out of the small windows on Salyut 1, and looked at the Earth below. The enormous area of the Soviet Union slowly drifted past, and he watched quietly, totally absorbed by the sight. He marveled at the fact that he was here at all. That his country was capable of producing a technological miracle such as Salyut 1 he had no doubt. However, without the succession of recent crew changes, his presence on this mission was most unlikely. He had certainly not thought that he would spend his 38th, and last, birthday in space.

In 1973, Owen Garriott spent a lot of his time looking at the Earth through Skylab's huge wardroom window. This window was the only one of note on the station, and to begin with, the stations designers had resisted including it, finally giving in to pressure from the potential crews. Now the crew could not imagine life without it. The work schedule aboard Skylab was intense, but each crewmember of the three missions tried to find some time each day just to look.

Georgi Grechko loved being back in space. He had flown to Salyut 4 two years earlier, in 1975, but the Salyut 6 station that he was now aboard was a great improvement in many ways. For one thing, it had a bigger, clearer window, and Grechko never tired of gazing at his homeland, and the far reaches of space. Many things had changed on the surface of his home planet in the time between the launch of Salyut 1 and now. Relations with the United States were more open then ever since the Apollo–Soyuz docking mission in 1975, and it was possible that more joint missions would take place in the future.

Ulf Merbold had trained for five years for the opportunity to fly aboard America's space shuttle, and now in 1983 he was here with his five crewmates aboard Columbia for the first flight of the European Spacelab. The schedule was unbelievably tight, but when he could steal a few moments, often before going to sleep, he would look at the Earth through the shuttle's flight-deck overhead rendezvous windows. Eleven years later, he would look again, but not through the windows of a space shuttle, but the windows of a Russian space station, called Mir.

The period that Michael Foale most enjoyed was when he had finished exercising. Hot and sweaty, he would float to one of the windows in Mir's Kristall module, this window was special because it had an air jet fitted that was originally used to cool a camera. The camera was long gone, but the jet remained and it was the ideal thing to cool down a steaming astronaut as he watched the world go by. Six years later Foale was looking through a much larger window than had ever been in space before. He floated in the U.S. Destiny laboratory module aboard the International Space Station (ISS) after exercising in the station's node module, Unity, and looked through the 20-inch-wide window at the Earth below.

Sergei Krikalev had flown in space six times: twice on the Mir space station, twice on the U.S. space shuttle, and as a member of the very first crew of the ISS. Now he commanded that station's eleventh expedition, and when this mission was complete, he would have flown over 800 days in space, more than any other human being. He had looked at the Earth from four different spacecraft, and once literally watched the world below change as the Soviet Union dissolved into the Confederation of Independent States before his very eyes. When he landed the communist state was no more, and he was a Russian citizen.

For Frank Culbertson it was the most painful experience of his life. Below him, the twin towers of the World Trade Center in New York lay in ruins, and every orbit allowed him to see the devastation from an unprecedented viewpoint on board the ISS. The Pentagon had been hit too, of course, and Frank was to learn that the pilot of that plane was a friend that he had been in flight school with. Tears don't flow as easily in space, he would later observe.

The history of man's space stations is a long one, and one that is necessary if we are to journey beyond the orbit of our own planet again. The glory days of Apollo are a long way behind us, many more manned hours aboard the ISS the space shuttles, and the Crew Exploration Vehicle (CEV) lay ahead before we can fulfil our destiny to land human explorers on Mars. Here is the story of what has gone before, the human story, the technical story, and the sometimes tragic tale of "The Story of Manned Space Stations".

1

1928–1970: How it all began

Life in the United States of America in the 1950s was pretty good. After the end of the Second World War, America was entering a Golden Age. The war effort which had provided tanks, planes, and ships, was now focused on providing more luxurious items to an eager population that may have only made up 5% of the world's total, but that was wealthier than the other 95% combined.

The only blot on the landscape was the Soviet Union. This was the McCarthy era, and the Senator from Wisconsin had made it very clear to all Americans that the enemy was without doubt Red. Most of his accusations were, in fact, totally groundless, but his point had been well made and remembered by the U.S. public. When the U.S.S.R. launched the first satellite, Sputnik, in October 1957, the paranoia that Joseph McCarthy had begun returned with full force. It suddenly seemed that America could not do anything right. When the U.S. responded with their attempt at a satellite launch in December, it exploded after achieving the heady heights of about two feet. They were finally successful in January 1958, but the other four launches that year also failed publicly, and there were many further very spectacular, very public spaceflight failures over the next decade. Meanwhile, it seemed like the Soviet Union could do nothing wrong, they seemed to enjoy success after success in the field of spaceflight, up to and including the flight of Yuri Gagarin, the first human into orbit, in April 1961. By comparison, the U.S.A. were not yet ready for a manned spaceflight, and just one week after Gagarin's flight, the U.S. suffered the Bay of Pigs invasion in Cuba that brought further embarrassment to the nation. The truth, of course, was a little different. In much later years we would learn that the U.S.S.R. suffered many failures in their space program, but this was not known at the time, and anyway the American public was not going to let a little thing like the facts get in the way of their opinion that somebody somewhere was sleeping on the job.

In 1961, therefore, the pressure was on new U.S. President John Kennedy to restore some pride to the nation, and if that sent a message to those pesky Russians at the same time, all the better. The question that Kennedy asked his advisors was,

"What can we beat the Russians at?" He was advised that simply trying to launch a space station ahead of their rivals would be a waste of time; the Soviets had already demonstrated that they had the lifting capability to achieve that goal before the U.S.A., and another "first" to the communists at this stage was unthinkable. So Kennedy's mind was made up for him, a month after Gagarin's flight, and with only 15 minutes of U.S. manned spaceflight experience behind him in the shape of Alan Shepard's ballistic flight, he announced the challenge of putting a man on the moon before the Soviet Union, and of doing so before the end of the decade. This was not to be a scientific endeavor, nor a noble crusade, it was to be a simple politically motivated challenge to the Russians to get there and back first, ideally without killing anyone in the process. It was not really what NASA wanted to do. The space agency knew that it was not ready for this, it had not even put a man into orbit yet, and now it was being asked to build the equipment needed to send men 250,000 miles to the moon and back, whilst at Cape Canaveral it seemed that every other rocket launch ended in a big bang. As we will see this crash program to send men to the moon and back did little to promote the cause of manned space stations, and in fact, simply got in the way of a logically progressive manned spaceflight effort. Not that Project Apollo and the Soviet moon program stopped all thinking about space stations, it did not, but it certainly meant that such ideas took a back seat to the preparations for landing a man on the moon.

America began to claw back the ground lost to the U.S.S.R. at the opening of the space race. In February 1962 Project Mercury put John Glenn into Earth orbit. In 1965–1966 Project Gemini, a two-man spacecraft, managed its own "firsts" in space-flight, and out-stripped the Russian space program, which was having problems of its own behind closed doors, in every area. And Project Apollo succeeded in landing men on the moon even before the Soviets were on the starting block. Meanwhile, others were thinking about space stations, all about a more permanent presence in space, where science and discovery were the motivating factors. Such thinking had begun many years earlier, almost as early as the dawn of flight itself.

1928—THE NOORDUNG STATION—HERMANN NOORDUNG

In 1928, Hermann Potocnik Noordung published his first and only book *Das Problem der Befahrung des Weltraums—der Raketen motor* (literally translated as *The problem of driving on space—the rocket engine*). This book was primarily concerned with manned space stations, the first in history to do so. It contained a design for a wheel-shaped structure for living quarters, with a power-generating station attached to one end of the central hub, and an astronomical observation station at the other end. He was among the first to suggest a wheel-shaped design for a space station in order to produce artificial gravity, and he pointed out the scientific value of such a station in a synchronous orbit above the Earth. His ideas were to inspire Hermann Oberth, and later Wernher von Braun and Sergei Korolev.

Sadly, Noordung himself did not profit from his amazing foresight, he died at the

Hermann Nordung, 1946

early age of 36 in great poverty, and his obituary in the local newspaper mentioned nothing of his spaceflight publication.

1946—THE VON BRAUN STATION—WERNHER VON BRAUN

In a 1946 summary of his work during World War II, Wernher von Braun prophesied the construction of space stations in orbit. The design, which owed a great deal to the earlier work of Noordung, consisted of a toroidal station spun to provide artificial gravity. Von Braun elaborated on this initial design at the First Symposium on Space Flight on 12 October 1951 hosted by the Hayden Planetarium in New York City. The design was popularised in 1953 in a series in *Colliers* magazine, illustrated with a gorgeous painting by Chesley Bonestell.

1948—THE BRITISH INTERPLANETARY SOCIETY STATION—H. E. ROSS

In a paper presented to the British Interplanetary Society (BIS), and reprinted in the Journal of the BIS in 1949, H. E. Ross described a manned satellite station in Earth orbit that would serve as an astronomical, zero-gravity, and vacuum research laboratory, and also serve as a way-station for the exploration of the moon. His suggested design comprised a circular structure that housed the crew of the space laboratory

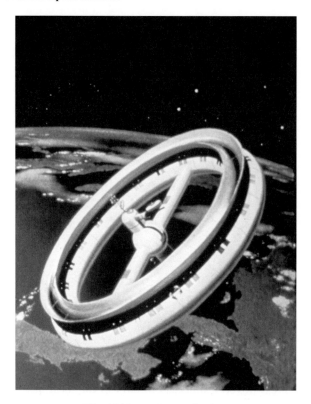

The 1946 von Braun Station

(numbering 24 specialists and support personnel) as well as telescopes and research equipment. The station, he suggested, could be resupplied with oxygen and other life-support essentials by supply ships launched every three months.

1954—EHRICKE FOUR-MAN ORBITAL STATION—KRAFFT EHRICKE

In "Analysis of Orbital Systems," a paper read at the fifth congress of the International Astronautical Federation in Innsbruck, Austria, Krafft Ehricke described an orbital station. Arguing that a very large space station was neither necessary nor desirable, Ehricke postulated a four-man design that might serve a number of different purposes, depending upon its altitude and orbital inclination. He suggested that such a station might be used for a variety of scientific research, for orbital reconnaissance, as an observation platform, and as a launch site for more distant space ventures. Later in 1958 Ehricke outlined the design for this station and called it Outpost. It would consist of an empty Atlas rocket equipped only with a pair of two-man gliders to serve as lifeboats, and could be powered by a nuclear reactor. Three further launches by Atlas-Centaur boosters would carry all of the remaining

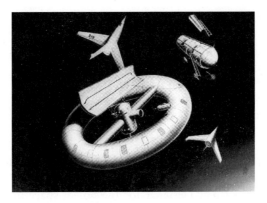

The 1948 BIS Station

equipment required by the station. The crews would also be launched by Atlas-Centaur. Future plans called for Outpost II and III, each of which would be bigger than the last, consisting of clustered Atlas boosters, with the whole station spinning to provide artificial gravity to the occupants in either end.

1959—THE PROJECT HORIZON STATION—WERNHER VON BRAUN

In 1959 Wernher von Braun and his team issued the Project Horizon report. This outlined the establishment of an entire lunar base by 1964. Von Braun at that time was with the Army Ballistic Missile Agency, and had yet to be transferred to the newly formed NASA. As part of the Horizon report, he advanced the theory that he had conceived years earlier for using a booster's spent stage as a space station's basic structure. The Earth orbital station was a major requirement for Project Horizon to succeed as there were no boosters on the drawing boards that could provide anything like the thrust needed to send the men and equipment for the lunar base to the moon under a direct ascent mode. An Earth orbit rendezvous would be required for refueling prior to flight to the moon. The "mode question" would of course later resurface when Project Apollo began. Project Horizon envisioned moving quickly to an early improved station constructed from 22 upper stage shells. Prior to any expansion of lunar outpost operations, sufficient tankage would have been placed in orbit to permit construction of two or three such stations. The orbital station crew strength was approximately 10; however, they would be rotated every several months. It was proposed in the report that the Earth-orbiting station created during the construction of the lunar outpost would continue as a separate program making use of the resources created rather than wasting them. The contributions that the space station would have provided were as follows:

- space laboratory, acclimatisation, and training capability for personnel;
- space laboratory for equipment;

- material storage space;
- low-altitude communication relay;
- Earth surveillance (perhaps a security consideration in this specific operation);
- space surveillance;
- meteorological surveillance;
- survey/geodesy data collection; and
- instrumentation for the test of Earth-to-space weapon effects.

1960—THE ORBITAL STATION (OS)—SERGEI KOROLEV

Sergei Korolev was the Chief Designer for the Soviet space program, although his identity did not become public until after his death in 1966. He was head of the OKB-1 design bureau that is now known as RKK Energia. In 1960 he made the first of many attempts to get the Soviet government to fund a manned space station as a logical progression of the fledgling manned spaceflight program.

On 23 June 1960 Korolev wrote to the Ministry of Defense in an effort to obtain support for a military Orbital Station (OS), on which a decision had been deferred to the end of the year. The station would have a crew of 3–5 and orbit at 350–400 km altitude. Its role would be to conduct military reconnaissance, control other spacecraft in orbit, and undertake basic space research. The first version of the station would have a mass of 25–30 tonnes and the second version 60–70 tonnes. Korolev pointed out that his design bureau had already completed a draft project in which 14 work brigades had participated, and so had a detailed plan.

1961—THE TOSZ STATION—SERGEI KOROLEV

The TOSZ—Heavy Orbital Station of the Earth—was Korolev's 1961 project for a large military space station. The draft project was completed on 3 May 1961, and marked the beginning of a long struggle throughout the 1960s to get such a station built and launched. Such a station required, of course, the N-1 rocket, the only rocket with anything like the payload lifting capacity required for such a large and heavy object.

1962—THE OS-1—SERGEI KOROLEV

Work on the OS-1 began on 25 September 1962. Following a meeting between President Nikita Khrushchev and the chief designers at Pitsunda, Khrushchev ordered that a 75-tonne manned platform with nuclear weapons be placed into low-Earth orbit (dubbed elsewhere as "Battlestar Khrushchev"). Korolev was authorized to proceed immediately to upgrade the three-stage N-1 vehicle to a maximum 75-tonne payload in order to launch the station. By 1965 the mock-up of the huge station had been completed. By 1969 the OS-1 had evolved to this configuration, as described in the official RKK Energia history. In 1991 engineers from Energia and other design bureaus taught a course on "Russian Manned Space" at the Massachusetts Institute of Technology (MIT). Dr. Vladimir Karrask, the first chief designer for the UR-500 (Proton), told of a shroud that he designed for the N-1. The shroud was cylindrical—6 m diameter × 30 m long—with a very "Proton-like" blunt conical top. He indicated that it had flown on the N-1. Another engineer, S. K. Shaevich, stated that flight hardware (including a back-up) was ready for the N-1 flights. There are those who believe that the last two N-1 flights had the Karrask shroud, and possibly the OS-1 station. It is not known if any OS-1 stations actually reached any stage of completion. Although plans for the OS-1 had to be constantly deferred until the N-1 booster proved itself, this did not prevent the design team from undertaking an even more grandiose study—the MKBS—in which OS-1 derived modules would form mere subunits of a huge space complex. At any rate the termination of the N-1 launch vehicle program ended any possibility of launching the station—unless it was reincarnated as the "Mir 2" jumbo space station that was planned for launch by the Energia booster in the 1990s.

1962—SOVIET N-1 LAUNCH VEHICLE PROGRAM BEGINS

The Soviets had long realised that in order to put many of their space projects into production they would require a heavy lift launch vehicle. Design studies had begun a few years earlier, but in September 1962 the official go ahead was given by the Central Committee of the Communist Party to begin the program in earnest.

The original design requirements for this giant rocket called for it to be capable of launching 75 tonnes of payload to orbit, and this dictated that the dimensions of the rocket were huge. It stood 344 feet tall, its first stage comprised 30 engines producing 43,000,000 kN of thrust, and it weighed 2,735 tonnes. The requirements were initially formed by the needs of the OS-1, but these requirements grew in the years before its first test launch in 1969. Building work began in March 1963 to create a complex of two launch pads for the giant rocket, and they were completed in 1967. The growing requirements of the Soviet lunar missions put continuous pressure on the already over-burdened N-1 design, and Korolev, and as of 1966 his successor Vasily Mishin, were forced to ask more and more of the stages and the engines that powered them. The N-1 was eventually to be capable of launching 95 tonnes; 20 more than originally specified.

1962—THE MODE DECISION FOR APOLLO—JOHN HOUBOLT

At first glance, the method for getting a man on the moon would not appear to have much relevance to the future of manned space stations, but in fact the method that was eventually chosen had a profound effect. The choices were simple enough, and there were three methods to be chosen from: Direct Ascent, Earth Orbit Rendezvous (EOR), or Lunar Orbit Rendezvous (LOR). There were others, but they were mostly far too risky/crazy to even be considered, the most ludicrous being the proposal that one man be launched to the moon where he would wait until such time that NASA figured out how to get him back!

Direct Ascent seemed the simplest of all: launch a giant rocket straight at the moon, without pausing in Earth orbit, land on the moon, and launch straight back to Earth. However, there were drawbacks, first, the rocket would be massive, far larger than the Saturn V that was eventually selected, and its launch facilities would be equally large and demanding. Second, the crew would launch at the top of this massive stack on their backs, as had been the case with Mercury and Gemini, which would mean that they would have to land on the moon in the same way; in other words they would not be able to see where they were landing. Various contraptions were devised to allow the crew (and their instrument panels) to swivel to an upright position for the lunar landing, but none seemed very practical, and of course the instrument panels would have to carry all of the information for lift-off, translunar coast, and landing, which would pose a daunting challenge to design and to operate. Third, imagine backing a vehicle of the size of an Atlas rocket down to the lunar surface; even if you managed it, you would be faced with a trek down a very long ladder to get to the surface.

The most supported mode initially was that of EOR; certainly it was supported strongly by the Marshall Space Flight Center under the directorship of von Braun. Cynics suggested that they supported this mode because it would need several rocket launches, and rockets were, of course, the responsibility of Marshall. This mode basically consisted of two or more launches into Earth orbit, where the moon bound vehicle would be assembled and fueled before setting off for the moon. It had the benefit that all rendezvous operations would take place in Earth orbit, allowing an immediate abort option. The downside was that it was more complicated due to its reliance on multiple launches, orbital rendezvous, dockings, assembly, and refuelling. However, EOR naturally included the option of building a staging post or space station to act as an assembly point for the moon ship; certainly this was favored by von Braun himself. Had this happened, our space station story may have had a very different beginning for NASA.

The third mode option, LOR, was a late comer, being first proposed in 1960 by a man named John Houbolt from the Langley Research Center to almost anyone who would listen to him, including all of the potential sub-contractors, and many within NASA. In June 1962 von Braun put his weight behind the LOR plan, and then in July NASA announced that would adopt LOR as the primary option for Project Apollo. It had the advantage that it would require only one launch vehicle which would contain all of the components required for the mission, and NASA chose the Saturn V to serve this role. The downside was that it required a rendezvous and docking in moon orbit where there was no abort option. Also the Lunar Module (LM) only had one ascent engine; if it failed, the crew of two would have a longer stay on the moon's surface than anticipated. However, the biggest advantage was that it could be implemented much more rapidly than the other two modes, and therefore get men on the moon within Kennedy's deadline. Nevertheless, LOR left NASA with nothing to build on. Had the challenge from Kennedy not

John Houbolt

arisen, there was a much more logical, albeit slower, way of getting man to the moon.

1963—MANNED ORBITING LABORATORY (MOL)—UNITED STATES DEPARTMENT OF DEFENSE

With NASA now enjoying great success with their manned spaceflight program, the U.S. Air Force wanted to be more involved. Their earlier project DynaSoar, which was to have been a manned orbital space plane, was in budgetary limbo, and NASA had not selected as many Air Force candidates for astronaut training as the top brass would have liked. Air Force manned space projects were not new: before Project Mercury it had created the "Man In Space Soonest" or MISS program, but this had been ignored when President Eisenhower decided in 1958 that he wanted manned spaceflight to be handled by a civilian agency, and established NASA.

In 1962, the USAF began to look closely at the proposed Gemini program, and realised that it held great potential to be modified for Air Force use, it had the added bonus that it would be tested first by NASA and it would be ready to fly much earlier than their own DynaSoar. With the addition of a cylindrical pressurized habitat that would be launched attached to the bottom of a modified Gemini spacecraft, the idea grew into MODS, or the Manned Orbital Development Station. However, even this interim project would not be ready early enough in the eyes of the Air Force brass, and it was proposed to fly a number of Gemini missions, in partnership with NASA, under the banner of "Blue Gemini". Unfortunately, the potential of this joint program was undermined when Secretary of Defense Robert McNamara demanded that not only should the Air Force take over the entire Gemini program, but all future low-Earth-orbit missions as well. NASA officials were naturally aghast at this prospect, and protested strongly that such a move would destroy America's plans to land on the moon by the end of the decade. Perhaps more surprisingly, senior USAF officers were similarly opposed to this plan, because they did not want their interim plans for a Blue Gemini, which they viewed purely as a means of gaining flight experience, to interfere with the larger DynaSoar project for which they had great hopes. Upon hearing of the Air Force's objections McNamara appeared to back down, and a new agreement was reached which merely allowed the installation of Air Force experiments on NASA Gemini flights. No sooner had this been agreed, McNamara took his revenge for the USAF's lack of support as he saw it, and cancelled both Project MODS and Blue Gemini. In fact these were just two of thirteen new projects for which the Air Force had sought funding in January 1963, and McNamara canceled them all, bringing to mind something about a secretary scorned! In December 1963 he rounded it all off by canceling DynaSoar as well. A bone was thrown to the Air Force, however, in the form of a new project known as the Manned Orbiting Laboratory (MOL). Essentially, MOL was MODS reborn. MOL was to be launched with its crew in a Gemini capsule, to be used once, and then discarded.

It was not until August 1965 that official funding for MOL came through when President Johnson allocated $1.5 billion to the program. With the program now in development, it was decided to begin the construction of launch facilities. MOL called for a base that could launch the vehicle into a polar orbit, a first for manned spaceflight, and so Vandenberg Air Force base on the California coast was chosen. Construction began in March 1966 of Space Launch Complex 6, or Slick 6 as it became known. The first real success of the MOL program came in November that same year when an already flown Gemini spacecraft, that had been modified to have a hatch installed in its heatshield, was launched not from Vandenberg but from Cape Canaveral atop of a Titan IIIC booster with a Titan II propellant tank standing in for the MOL beneath the capsule. The capsule was successfully recovered and represented the first reusable spacecraft launch and recovery. However, by the end of that first year, despite continuing progress, the program was faltering under the load of ever increasing costs, and a falling budget. Also becoming a problem was the ever increasing weight of the MOL combination, which in turn required the man-rated version of the Titan IIIC, known as the Titan IIIM, to be upgraded with additional segments to its solid rocket boosters. However, despite the program's difficulties, progress was being made, Slick 6 was nearing completion, and the Air Force had recruited 14 astronauts.

The MOL pilots were recruited in three groups in much the same way that NASA appointed its astronauts. The first group, which was chosen in 1965, consisted of eight pilots; six from the U.S. Air Force, and perhaps surprisingly, two from the U.S. Navy. They were:

Michael J. Adams, USAF
Albert H. Crews, USAF
John L. Finley, USN
Richard E. Lawyer, USAF
Lachlan Macleay, USAF
Francis G. Neubeck, USAF
James M. Taylor, USAF
Richard H. Truly, USN

This group was different from the NASA astronaut selections in that they were all active military, and were all pilots, a moniker that they retained rather than calling themselves astronauts. All were handpicked from a list of Aerospace Research Pilot School (ARPS) students, instructors, and graduates by Chuck Yeager, the ARPS commandant.

The second group were selected the following year, and consisted of five more pilots:

Robert F. (Bob) Overmyer, USMC
Henry W. (Hank) Hartsfield, USAF
Robert L. Crippen, USN
Karol J. Bobko, USAF
Charles Gordon Fullerton, USAF

Again, one year later, in 1967, a third and final group was chosen, only four pilots this time:

Robert T. Herres, USAF
Robert H. Lawrence, Jr., USAF
Dr. Donald H. Peterson, USAF
James A. Abrahamson, USAF

First MOL astronaut selection

Second MOL astronaut selection

Third MOL astronaut selection

Of these fourteen pilots nearly all would go on to continue their careers with some distinction. From the first group, Mike Adams left the MOL program and transferred to the USAF X-15 program where he successfully completed six flights before he was killed on his seventh flight after the aircraft experienced technical difficulties that put it into a spin at about 206,000 ft. Adams recovered from this spin, but the aircraft disintegrated under the 15 g loads and fell from the sky. Richard Lawyer left and had a distinguished test pilot career, before re-entering the MOL story toward the end of his life as we will discuss later. Robert Lawrence would have been the first African American in space, but he was killed in 1967 whilst flying in the backseat of an F104 on an ARPS mission to practise X-15-type landing approaches. The pilot of the aircraft, Major Harvey Royer, misjudged his approach and hit the runway hard, the undercarriage collapsing and launching the aircraft back into the air, now ablaze at its rear. It landed 2,000 ft further down the runway and disintegrated as it bounced once more. Both pilots ejected successfully, but Lawrence's parachute failed to deploy, and he was killed—Major Royer survived the accident.

The MOL program received a shock in June 1969. All of the major program officials fully realized that the program was late and over budget, and they certainly feared that when the budgets were announced that would be left short, but it seems that no-one had actually considered complete cancellation of the project, which is what President Richard Nixon did. The MOL pilots were offered the opportunity to transfer to NASA. Seven took up the offer, and all went on to become important members of the space shuttle program. Richard Truly flew STS-2 and STS-8 before retiring from NASA, and later became the 8th NASA Administrator between 1989 and 1992. Robert Crippen was the pilot of the very first space shuttle flight aboard Columbia in 1981, and went on to fly two more missions as shuttle commander. Ironically he was to command the first mission of the space shuttle from the same Vandenberg Slick 6 launch complex that MOL was to have launched from, but that mission was canceled after the Challenger accident, and Crippen retired from NASA. He too later returned in a management capacity, acting as Director of the Johnson Space Center between 1992 and 1995. Karol Bobko also flew the shuttle three times, once as pilot and twice as commander. Bobko is still involved in the spaceflight business as Vice President of SpaceHab. Gordon Fullerton still works for NASA as a research pilot. He flew Enterprise in the Approach and Landing tests, and later on STS-3 and STS-51F, the latter a Spacelab mission which to this date holds the distinction of being the only in-flight abort of the shuttle program. Henry Hartsfield was another to fly the shuttle three times, once as pilot on STS-4, and twice when he commanded STS-41D, the first flight of Discovery, and STS-61A, a Spacelab mission that was the first in history to have an eight-person crew. Unfortunately, Bob Overmeyer was killed in March 1996 whilst piloting an aerobatic aircraft; but he had successfully flown two shuttle flights, his first as pilot on board the first operational shuttle flight, STS-5, and the second in command of flight STS-51B, a Spacelab mission. Don Peterson flew just one mission on the space shuttle, STS-6, and was the only member of his group not to fly as pilot but as mission specialist. However, his mission specialist designation allowed him also to be the only member of the group to carry out a spacewalk.

Colonel Richard Lawyer re-enters our story in June of 2005, when artefacts from the abandoned MOL program were found at Cape Canaveral Air Force Station in Florida. A room at the launch complex 5/6 museum that had apparently not been opened for many years was being checked by security officers when two blue MOL spacesuits that had been used for training were found to be in almost perfect condition, one belonged to Lt. Col. Richard Lawyer. Other MOL spacesuits are on display at the USAF Museum in Dayton Ohio, and at the Johnson Space Center. The two newly discovered suits were donated to the Smithsonian Institution. Unfortunately Colonel Lawyer died later that same year. To the very end he had kept his vow to keep his country's secrets. While very forthcoming about general aspects of the MOL program, he would never say a word about its actual mission. He would simply say, "I am not at liberty to deny or confirm the reported mission for MOL."

1964—BIRTH OF ALMAZ AND SALYUT—SERGEI KOROLEV AND VLADIMIR CHELOMEI

On the Soviet side it was all about the competition between two implacable rivals; Sergei Korolev head of the OKB-1 design bureau and responsible for all of the Soviet Unions space successes so far, and Vladimir Chelomei, head of the OKB-52 design bureau, which had a great deal of experience with missiles, but no track record in space. Korolev had been tasked with developing the Soviet lunar program in order to compete directly with NASA. Chelomei, who had the support of the military, was designing a manned surveillance platform, which he called Almaz, to be serviced by a manned ferry/cargo craft called the TKS. The crew of three would be launched with the Almaz station aboard a returnable capsule, gaining entry to the station via a hatch in the heat shield. They would be launched with as much food and water as possible, but at some point a TKS would be flown to a docking by another crew (automatic dockings had not yet been developed) to facilitate resupply and crew exchange. Chelomei's design, whilst certainly innovative, and more flexible than the USAF MOL, suffered from his own and his bureau's lack of real spacecraft experience, and soon fell far behind schedule.

Korolev, however, was having his own problems with his new Soyuz spacecraft design. The first Soyuz launch was rushed before it was really ready, culminating in April 1967 with the death of cosmonaut Vladimir Komarov. But Korolev did not live to see this. He died in January 1966 during a routine operation. Vasily Mishin had the unenviable task of replacing Korolev, and his task was not helped by the fact that he and Chelomei hated one another to the point that they could not stand to be in the same room together, making collaboration or co-operation virtually impossible.

1966—APOLLO APPLICATIONS PROGRAM (AAP)—NASA

When George Mueller took over as director of NASA's office of manned space flight in 1963 he set out to ensure that after Apollo had achieved the first lunar landing, the tremendous technical capability developed to achieve this feat should not be wasted. So was born the Apollo Applications Program, and in March of 1966 the first AAP schedule was revealed. It was adventurous to say the least. It projected 45 launches using both the Saturn V and Saturn IB to both Earth and lunar orbits, all of these missions separate from the moon landing effort of Project Apollo. Most significantly, these launches included three Saturn S-IVB Spent Stage Experiment Support Modules (SSESM), otherwise known as "wet workshops". This form of space station seemed an economical way for NASA to obtain its first space station experience. The S-IVB stage would be launched to orbit in the normal way as the upper stage of a Saturn V, with a crew in an Apollo CSM, but the spent stage would remain in orbit where it would be dried out internally and outfitted by the crew as a temporary laboratory and workshop. There were some concerns within NASA over this approach, not least within the Astronaut Office, which was primarily concerned with the suitability of a emptied hydrogen tank for human habitation, plus the issues of providing power to the planned experiments, and the general safety of such a structure.

In November 1967 the Manned Spacecraft Center proposed an alternative to the "wet workshop", a "dry workshop". This basically meant that instead of launching the S-IVB stage as an active part of the booster and then outfitting it in orbit, the stage should be outfitted on the ground and launched as a conventional payload. However, there was some opposition to this proposal, and it was decided to continue with the wet workshop plan. Things changed again in May 1969; the early success in man-rating the Saturn V had potentially freed up a Saturn V. This reopened the dry workshop possibility. The benefits of being able to completely outfit the workshop on the ground before launch were clear, and Wernher von Braun and his team at Marshall began to warm to the idea that they had originally opposed. In June of that same year, the Department of Defense MOL program was canceled, and several elements including seven of the program's astronauts, were transferred to NASA. This added new momentum to the Orbital Workshop Program (OWS), as the sole-remaining element of AAP had become known. In July 1969 Apollo 11 landed on the moon, and NASA's Administrator, Tom Paine, approved the change from wet to dry workshop design, and officially assigned a Saturn V to launch it. The number of AAP launches had now reduced dramatically to just four: one Saturn V to launch the workshop, and three Saturn IB launches to get the crews to the orbiting outpost. In February 1970, the project received an official name; America's first manned space station would be called Skylab.

1969—SOYUZ 4 AND 5—FIRST DOCKING BETWEEN MANNED SPACECRAFT

Early in 1969, the Soviets laid claim to having formed the first space laboratory with the docking of two manned spacecraft, the first in history. Soyuz 4 with Vladimir Shatalov on board was launched on 14 January. Soyuz 5 was to follow 24 hours later, the delay being to allow time for Shatalov to acclimatize to orbital conditions before attempting the docking. Soyuz 5 had a crew of three, commander Boris Volynov, and flight engineers Aleksei Yeliseyez and Yevgeni Khrunov. The docking was handled manually by both commanders, and was achieved flawlessly; the first docking of two manned spacecraft. However, appearances can be deceptive. Whilst the two space-craft were physically docked together, it was not possible for the crews to float through the hatches on the nose of the connected orbital modules. Transfer between the two spacecraft was only possible by way of Extravehicular Activity (EVA), or "spacewalking" as it is more popularly known, making use of the side hatches in the orbital modules of both craft. This EVA was a necessary test of the method that

Shatalov describes Soyuz 4 and 5 docking

would be used by a cosmonaut during a Soviet Moon-landing mission. The Soviet lander also had no internal hatch to allow transfer between the vehicles. Yeliseyez and Khrunov carried out the EVA successfully and returned to Earth with Shatalov on board Soyuz 4. Volynov returned with Soyuz 5 alone. Volynov's re-entry was not without incident, however. He failed to orientate his spacecraft prior to entry, and to add to his problems the propulsion module had not separated completely from his descent module, which caused the spacecraft to tumble and face the wrong way for re-entry. Just as Volynov thought that disaster was near, the module separated, and his descent module turned to face the right way. His problems were not over yet; the parachute lines then began to tangle, but fortunately sorted themselves out before the parachutes had fully inflated, and he landed successfully, although much harder than normal, he broke free from his harness, and broke several of his front teeth against the opposite bulkhead. He staggered from his capsule and found a peasant's hut where the occupants cared for him until help arrived; he was grounded for two years.

It had not been a long duration flight by any means, and the spacecraft remained docked for only four and a half hours, but it had been a successful prelude to manned dockings with orbital space stations, if not perhaps the world's first space laboratory.

1969—FIRST TEST LAUNCH OF N-1

In February 1969, the first test launch of the N-1 ended in disaster. The rocket was in trouble immediately after its lift-off: a fire had developed in its first stage that grew worse as the rocket ascended, and when the engine-monitoring system detected the fire 68 seconds into the flight, it unfortunately responded by shutting down the entire first stage, and the enormous vehicle crashed back to the ground. The N-1 program, which had been in trouble since its inception, had floundered. The Soviet hierarchy realized that any chance of beating the U.S.A. to the moon had crashed along with this first test flight.

1970—SOYUZ 9—LONG-DURATION FLIGHT TO BEAT GEMINI 7

The sole objective of the flight of Soyuz 9 was to set a new spaceflight endurance record, and beat the previous best of fourteen days that had been set by Gemini 7 five years previously. The crew consisted of commander Andrian Nikolayev and flight engineer Vitali Sevastyanov. Nikolayev had previously flown on Vostok 3, and he was married to cosmonaut Valentina Tereshkova. Sevastyanov was making his first flight. The Soyuz had been specially modified to undertake this long-endurance flight: its docking system had been removed, and a new larger life support system had been installed. The already cramped orbital module had also been fitted with exercise equipment and extra storage racks, as well as additional carbon dioxide scrubbers. The crew launched successfully on 1 June 1970, and immediately started work on their extensive suite of scientific experiments. Unfortunately, they devoted so much of their time to experiments that they neglected their physical exercise program, with the result that when they landed eighteen days later they were unable to stand and took several weeks to recover fully. Of course, the flight was not just about testing the ability of the human body to withstand weightlessness over an extended period, it was equally important that the Soyuz spacecraft prove itself to be capable of long stays in orbit because if it was to progress to acting as a ferry between the ground and an orbiting space station, it would have to remain in space for long periods. With the mission successfully completed, confidence in the Soyuz design was boosted. However, there was still much to learn about long-duration flight if cosmonauts on missions to space stations were to avoid the pitfalls of the Soyuz 9 crew.

1970—BORN OUT OF CHAOS—SALYUT—SOVIET GOVERNMENT

After the disasters of the Soyuz 1 and N-1, and the continuing disagreements between Mishin and Chelomei, the Soviet government decided that the rival teams should pool resources, under the program name DOS, in order, finally, to get the space station project off the ground. The basic Almaz design was thought to be sound and was kept, but the TKS ferry was thought to be too complicated for rapid development, and so modifications were made to the Almaz design to allow it to accept a Soyuz as the crew ferry. Other changes included replacing Chelomei's design for propulsion with the proven Soyuz engine module,

The result of this enforced collaboration was Salyut 1, the first chapter in the story of manned space stations.

2

1971: Salyut 1—triumph and disaster

The successful launch of Salyut 1 on 19 April 1971 was a truly historic event. Salyut had not always been its name, indeed the word Zarya was written on the side of the station. Just before launch the official name became Salyut, apparently to prevent confusion with a ground station already named Zarya. After so many years of dreams and plans, humankind had an orbiting space station, and it was ready to accept its first crew. The launch was particularly noteworthy for the Soviets as it came a full two years before America could launch its planned Skylab. This had been one of the main motivations behind combining the Almaz design with Korolev's Soyuz ferry vehicle. As with many of the Soviet's spaceflight achievements, political considerations had pushed the space station program forward faster than it might have on its own. This first station was not huge, weighing about 18 tonnes and measuring 20 m in length, and certainly not luxurious, but it represented a milestone in manned space exploration.

The crew of Soyuz 10 would be the first to inhabit this new outpost in orbit. The crew comprised commander Vladimir Shatalov, flight engineer Aleksei Yeliseyez, and researcher Nikolai Rukavishnikov. They were launched four days after Salyut 1 had successfully made orbit, and rendezvoused with the station shortly after. The docking was carried out without any problems, Shatalov having exploited his previous experience of docking Soyuz 4 and 5. Unfortunately, despite a hard docking having been achieved, the crew were unable to swing back the Soyuz docking probe that had to be removed before the crew could access the tunnel that joined the two craft. It was later determined that a failure in the Soyuz docking port's electrical system had caused the problem. The crew of Soyuz 10 had no choice but to undock from the station and return home, having filmed the Salyut docking port for later analysis on the ground.

The back-up crew for Soyuz 10 consisted of commander Alexei Leonov, with flight engineers Valeri Kubasov and Pyotr Kolodin, and they were now advanced to the prime crew for Soyuz 11. For Leonov this was a significant event. In the three

Soyuz 10 back-up crew

Soyuz 11 crew

years following the historic Voskhod 2 flight that had made him the first human to walk in space, he had been training for a flight around the moon in a Zond spacecraft. The flight of Apollo 8 in lunar orbit in December 1968 and a less than successful unmanned test of Zond, had led to his flight being canceled. Ultimately, the entire Soviet manned lunar program was canceled, and Leonov was promoted to lead the training of cosmonauts for the Salyut program. However, fate was to intervene in Leonov's career once more when Kubasov developed a lung infection shortly before launch. This was later determined to simply be an allergic reaction, but that did not help Kubasov at the time; he was removed from the Soyuz 11 crew and replaced with Vladimir Volkov, his back-up. Then, just eleven hours before launch, it was decided to replace the entire Soyuz 11 crew as a precaution against Kubasov's lung infection having been passed on to the rest of the them. Leonov was replaced by Gyorgy Dobrovolsky, and Kolodin by Viktor Patsayev. Volkov remained on the crew. The replacement crew for Soyuz 11 were as shocked as Leonov by the decision. They had only been training together for a few months, and had not expected to be launched on an actual mission for several more months, and were concerned that they were not ready. Leonov's crew were sent away for a holiday before they began training for a flight to Salyut 1 upon the return of Soyuz 11.

The launch, rendezvous, and docking of Soyuz 11 all went smoothly, and the crew were able to enter the station with none of the problems that had affected the previous flight. Despite their concerns and relative lack of training, the flight proceeded well for 12 days until 18 June when the smell of burning was detected and a small electrical fire was found. The crew were very alarmed by this and urged the ground controllers to let them evacuate the station and return to Earth. In preparation, they powered up the Soyuz ferry vehicle, but returned to the station when it was realized that the danger had passed. Nevertheless, this incident had badly dented their morale, and although they continued their work, it was with less passion and drive than before. After a week, the ground controllers decided to let the crew come home early, and on 29 June they packed the Soyuz for the return trip. Their mood was significantly lifted as they strapped themselves in and undocked from Salyut 1, thereby bringing to an end the first mission to a manned space station, which had originally been planned to last 30 days, but was cut short to 23 days.

The Soyuz re-entered the atmosphere as expected and parachuted to a soft landing on the steppes of Kazakhstan. The recovery team opened the hatch to find all three men dead in their couches.

The Soviet people were horrified by the deaths of three brave men that they had come to know well from their nightly broadcasts from the Salyut station, and they now mourned their loss as they would a family member. The crew were interred in the Kremlin wall alongside other space heroes such as Yuri Gagarin, Sergei Korolev, and Vladimir Komarov. The inquest soon determined that a pressure relief valve designed to equalize the internal pressure in the capsule as it descended through the atmosphere had opened prematurely, possibly when the explosive bolts that separated the descent module from the orbital and propulsion modules were fired after the de-orbit burn, prior to entry into the atmosphere. It would probably have not been immediately apparent to the cosmonauts that the valve had opened; and even if it had, the

Soyuz 11 undocks from Salyut 1 (computer image)

valve was not easily accessible by the crew, although there was evidence to suggest that they had tried to stem the flow of air from their craft. This failure would not have been a problem except for one important fact, the crew did not have pressure suits; Soyuz crews simply wore flight overalls. As the pressure inside their capsule vented, the crew slowly lost consciousness, and eventually died from embolisms in the blood due to the vacuum. The Soyuz landed automatically as if nothing was wrong. Alarm bells rang throughout the spaceflight community. NASA even contacted the Soviets to determine if the long duration of their mission had been a factor in their deaths. Clearly, changes needed to be made to the Soyuz design to prevent a future catastrophe, and Salyut 1 would not be able to be inhabited in its lifetime again, so it was commanded to de-orbit by firing its engines to initiate a ditching in the Pacific Ocean in October 1971.

The redesign of the Soyuz spacecraft turned out to be substantial. It was clear that in the future cosmonauts must launch and land wearing pressure suits, and this would require more room than was currently available in the descent module. The only way to accommodate the newly designed Sokol K1 spacesuits, along with the extra equipment needed to support the space-suited crew would be to remove one man from the Soyuz configuration. This had implications for future space station designs, as a crew of two would obviously have more work to do. The Soyuz 11 crew had spent much of their 23 days aboard Salyut 1 simply looking after the station's systems; two men would be even more pressed to keep up with a station's needs.

Alexei Leonov was assigned to command the first crew to occupy the next Salyut station, along with Valeri Kubasov, but his luck was to betray him again. The next Salyut was actually the back-up for the Salyut 1 mission, and therefore identical to its predecessor. Unfortunately, only two and a half minutes after launch on 29 July 1972 one engine on the Proton rocket's second stage failed, and the vehicle crashed into the Pacific Ocean, taking the Salyut with it. Officially it was never called a Salyut or anything else; only in later years would it become apparent that this launch had taken place.

3

1970–1979: Skylab—NASA dips its toe

In March 1970, the Skylab project received official approval by President Nixon when he referred to it during a speech about America's goals in space for the coming decade and beyond. However, this was a difficult time for NASA, they had achieved President Kennedy's challenge of landing a man on the moon before 1970, indeed they had done it twice with Apollo 11 and 12, and now they faced the inevitable post-success anticlimax, and the people of the United States lost interest. The Soviet threat to the moon landings had failed to materialize, and the risks of further moon landings were all too clearly demonstrated during the flight of Apollo 13 in April 1970. NASA's budget had been slowly reducing for years now, and finally they had to cut flights: two Apollo missions were deleted from the program that would now end with Apollo 17 in 1972. It was at this time that the first hint of co-operation with the Soviets became apparent, with a suggested docking of a Soyuz with the Skylab workshop. This was at a time, of course, when the Soviet's plans for their Salyut stations was completely unknown to the Americans until Salyut 1's launch in 1971. NASA then suggested that perhaps an Apollo CSM could dock with a Salyut station, but the Soviets were not keen on this idea, and NASA had already decided that a Soyuz docking with Skylab was also not an option any longer. These discussions continued, and eventually an Apollo–Soyuz docking was suggested, and this would lead to the Apollo–Soyuz Test Project (ASTP) of 1975.

In 1971 Chief of Flight Crew Operations, Deke Slayton, began the process of selecting crews for the upcoming Skylab missions. At that time, three missions were definitely scheduled with the possibility of two more. It had also been suggested that the crews should consist of one pilot/commander, preferably a flight experienced astronaut joined by two scientist-astronauts in order to maximize the scientific output from these flights. Slayton quickly put a stop to that idea; his feeling was that Skylab was a totally new kind of mission, and he wanted two pilot astronauts on each crew in case something went wrong. He came up with the following crew assignments based on those criteria.

Mission		Commander	Pilot	Science-pilot
Skylab 1	Prime	Pete Conrad	Paul Weitz	Joe Kerwin
	Back-up	Rusty Schweickart	Bruce McCandless	Story Musgrave
Skylab 2	Prime	Al Bean	Jack Lousma	Owen Garriott
	Back-up	Vance Brand	Don Lind	Bill Lenoir
Skylab 3	Prime	Gerry Carr	Bill Pogue	Ed Gibson
	Back-up	Vance Brand	Don Lind	Bill Lenoir
Skylab Rescue	Prime	Vance Brand	Don Lind	

Skylab 3 and 4 back-up crew

Even these initial assignments had undergone some change. Walt Cunningham had originally been assigned as back-up commander for the first flight, but he choose to leave NASA rather than stick around for another two years as only a back-up. He was replaced by Rusty Schweickart, who in turn was replaced on the Skylab 2 and 3 back-up crews by Vance Brand. Also added at a later date was the possibility of a Skylab Rescue mission. This was the first time that planning a rescue mission had even been possible in NASA's space program. It involved flying a special Apollo Command and Service module fitted with two extra couches underneath the outer-most couches already installed; this was a small area that had been used as a sleeping space during Apollo moon missions. This modified CSM would be flown by a crew of two, and come back with five crewmembers after docking with the second port on Skylab.

It was at this point that some confusion entered the Astronaut Office concerning the design of the mission patches for Skylab. The official designation for the three manned flights was SL-2, SL-3, and SL-4, with the first unmanned launch of the lab itself designated SL-1. The crews had designed their patches according to this numbering, but were later informed by the Skylab Program Director that in fact their flights were being referred to as Skylab 1, 2, and 3, so the patches were changed. When the patches were submitted for official approval, they were rejected by NASA's Associate Administrator for Manned Spaceflight, Dale Myers, because of their numbering, and he ordered them to revert to the original designations. However, it was too late for the crews to do this, as their clothing for their upcoming missions had already been stored on board Skylab ready for its launch. It was deemed far too expensive, and unnecessary to change the clothing and labels at this late stage, so although the office designations for the missions remained, the patches are labeled, 1, 2, and 3. Such are the difficulties of managing a space program!

With the flight crews and launch dates now defined, some modifications were required to the launch pads to support the launch of the Saturn IB rocket. This had been used only once previously for a manned launch, when Pad 34 had been used for the Apollo 7 mission. As that pad was no longer available, it was decided to modify Pad 39B to accept the Saturn IB, and leave 39A largely as it was to launch the last ever Saturn V booster with the Skylab workshop on board. Given that most of the upper connections on the much shorter Saturn IB were the same as for the Saturn V that Pad 39B had been designed for, it was decided that the easiest modification to the pad would be to build a 127 foot high pedestal for the Saturn IB to sit on. This pedestal became known as the milkstool.

The Skylab workshop itself had undertaken quite a journey. Built originally as the second stage of the Saturn IB launch vehicle, it now had to be converted into a useable orbital workshop. S-IVB second stage number 212 had been built in 1966 by McDonnell Douglas, and its accompanying J-2 rocket engine built and tested during 1967 and then installed into stage 212 later that same year. At that point in time this stage was not assigned to a specific mission, so it was put into storage at McDonnell's Huntingdon Beach assembly plant until March 1969. At the end of this period it was identified as being ideal for refurbishment as the Skylab orbital workshop. As 1969 progressed, the J-2 engine, thrust structures, and various other parts were removed to

Skylab

leave the stage consisted only of its two fuel tanks. It took a further two years of work to prepare the interior of the hydrogen tank for human habitation in space. The second smaller tank, originally intended for liquid oxygen, would be used by the crew for storing all of their trash. By the end of 1972 the Saturn S-IVB stage 212 was ready to be launched as the primary Skylab workshop. At the same time, another S-IVB stage, number 515, this time from a Saturn V, had been identified as the back-up Orbital Workshop and had gone through the same conversion process as stage 212. It never flew, of course, and it was delivered to the Smithsonian Institution for display at the Air & Space Museum in Washington D.C., where it has been since July 1976.

Before any of the announced crews could visit the station, it was decided to run a full mission length simulation on the ground. This simulation would allow all the experiments and equipment aboard the station to be tested before launch. It would also help to alleviate any medical fears regarding the crew's long-term exposure to a artificial closed ecological system. If there were any problems, it would be better that they happened first on the ground. In order to run the simulation as accurately as possible, a complete mock-up of the Skylab interior had to be built in an altitude chamber in order that the correct pressure and mixture of gases could be used. It was

SMEAT crew

decided to use the 20 foot diameter chamber at the Manned Spacecraft Center in Houston. The program was known as SMEAT, which stood for Skylab Medical Experiments Altitude Test. Originally planned to consist of two simulations, one lasting for 28 days and a second lasting for 56 days, it was decided to limit the program to just one 56-day test. The crew for the SMEAT test was to be selected from the pool of existing astronauts, but not to include any of the selected Skylab crews, their back-ups or support crew. Bob Crippen was selected as commander, with Karol Bobko as pilot and Bill Thornton as science-pilot. They designed their own mission patch, which featured the cartoon character Snoopy with a tightrope around his neck; this was said to reflect how they felt about some the medical experiments that were to be performed on them.

SMEAT patch

On the 26th July 1972 the three men prepared to start their marathon simulation with a medical check before beginning a long pre-breathing period to purge nitrogen from their blood. During the "mission" the crew participated in all the experiments that the actual crews would perform in flight. This allowed them to discover any problems with procedures, and to set a baseline for the experiments that were to be performed in orbit. The test ended on 20 September 1972, and undoubtedly made a massive contribution to the success of the Skylab missions.

By the end of 1972 the Skylab program was ready for its first launch. The thirteenth and final Saturn V booster to be launched would be used to haul the Orbital Workshop into space, where it would be visited and lived in by three separate crews launched by Saturn IB boosters from an adjacent pad. With the crew of the first mission watching, the Saturn V lifted itself from Pad 39A, and at first, everything appeared to be quite normal.

Unfortunately just as the vehicle was passing through Max Q (a term for maximum aerodynamic pressure) about 70 seconds after launch, the first signs from telemetry showed that the booster was in trouble. The telemetry showed that the micrometeoroid shield and the number two solar array had already been deployed. This, of course, should have been impossible, for the Skylab workshop was still surrounded by the aerodynamic launch shroud. In fact, the shroud enclosed only the structures atop the OWS. The skin of the OWS was the S-IVB, which was exposed to the airflow. However, the Saturn V continued its pre-programmed path and delivered Skylab to orbit. It now remained to be seen what condition the lab was in. Initial telemetry suggested that there had been a major problem with the solar arrays, as the amount of power being generated by them was a small percentage of what it should have been. Clearly if the station could not generate enough power, it could not be occupied for any length of time. After more detailed investigation by NASA officials, it was determined that a design imperfection had caused the micro-meteoroid shield to move away from its flush location against the workshop, and aerodynamic forces had then ripped the entire shield away, taking the left-hand solar array with it. It was uncertain whether the right-hand array had been similarly lost, or was trapped against the lab by debris from the departing shield. It was hoped that the

Skylab ready for launch

latter was the case. Pete Conrad's crew were stood down until they could be trained to free the trapped array. Unfortunately, the micrometeoroid shield was to have served as the thermal shield to keep the interior of the workshop cool. With its demise, the internal temperature was climbing steadily to the point where it would exceed the limits designated safe for human habitability. The obvious thing for ground controllers to do was to maneuver Skylab such that the area of bare skin was pointed away from the Sun in order to keep the internal temperatures under control. However, this also meant facing the remaining solar arrays, which were located on the Apollo Telescope Mount, away from the Sun, thus depriving the fledgling station of power. Eventually the Skylab controllers alternated the station between different attitudes in an effort to find the best compromise. A further complication caused by the increasing internal temperatures was the condition of the food supplies aboard the station for all three of its future crews. The temperature had risen to 54°C but it was determined that all of the canned food on board would survive such temperatures for quite a while if necessary. Further concerns affected the medical supplies and film—it was decided that the crews would carry fresh supplies with them.

Ultimately, however, it would fall to the first crew to make repairs to the station if the entire planned program was to be carried out. Many possible solutions for both the shield and the solar array problems were put forward, but most were not practical. Eventually 10 solutions were shortlisted, and after further deliberations this list was cut to two. It was decided to supply the first crew with both solutions. An improved Sun shield solution would be made ready for the second crew to install after the first crew had reported on the condition of the station. Testing of the components to be used by Conrad's crew was carried out by Schweickart and Kerwin in the neutral buoyancy water tank at the Marshall Space Flight Center, to develop proceedures and verify that the equipment would function as anticipated. The Extra-vehicular Activities (EVAs) planned for Conrad's crew were arguably the most complex, and the requirement to undertake them so early in the mission by a relatively untrained crew was greeted with nervousness by many within NASA. A simpler method for deploying a replacement temporary Sun shield was therefore devised that would enable the crew to remain inside the workshop, but for the stuck solar array there was no choice but to proceed with the planned EVA. The command module for the first crew would therefore be crammed with improvised and off-the-shelf tools to aid in the freeing of the remaining solar array.

Pete Conrad and his crew lifted off from the milkstool on Pad 39B on 25 May 1973, their destination the damaged Skylab Orbital Workshop. The rendezvous proceeded normally, and the first order of business was to fly around the workshop to carry out a visual inspection of the damage. After first docking with Skylab in order to conserve station-keeping fuel, the crew undocked to carry out a stand-up EVA. Conrad drew the command module up to the damaged solar array for a closer inspection; which revealed that a couple of metal straps were preventing the still intact array from deploying. They depressurized the command module and Paul Weitz and Joe Kerwin prepared to attempt to free the trapped wing. The procedure was for Kerwin to remain in the hatch and hold on to the legs of Weitz, who was hanging out of the hatch with a long-handled cutting tool. Every time Weitz

Skylab 2 crew

attempted to cut the metal straps he would inadvertently pull the command module nearer to the hull of the Skylab, which meant Conrad at the controls had to fire thrusters to prevent a collision, which in turn made Kerwin's task difficult. It just was not going to work. The crew now attempted to dock their spacecraft with Skylab's axial docking port again, but this time they had trouble, only completing a successful docking after they had disassembled the command module's docking mechanism and carried out repairs. Mission Control decided that this would be a good time for the crew to have a meal and a sleep period before entering the station.

When the crew did enter Skylab the next day, they found the temperatures to be extreme, about 125°F; Conrad likened it to the engine room on an aircraft carrier. Entering the workshop in short shifts and returning to the command module to cool off, the crew set about deploying the makeshift parasol. Making use of a small scientific airlock in the wall of the workshop on the sunward-facing side, they deployed the temporary sunshade in the fashion of a chimney cleaner extending his brush by adding a new section of rod and pushing it further up the chimney. Conrad and Weitz carried out the deployment, whilst Kerwin watched their progress from the command module. Once the parasol had been fully extended, it began to flatten itself in the warmth of the Sun, and soon the temperatures in the workshop

View of Skylab from Skylab 2 CSM

began to drop; although it took about a week for the temperature to drop below 70°F. The workshop was now habitable, and the crew moved their belongings into their individual cabins and began to unpack the contents of the station in preparation for carrying out their assigned scientific duties.

Power, however, was a big problem; with only the solar arrays on the telescope mount available, Skylab had less than half the power it required. The crew would have to venture outside and attempt once more to free the trapped solar wing. Conrad and Kerwin ventured outside with the various tools that had been loaded on board their command module. One tool was a very-long-handled cutter of the type used by telephone repair men to remove branches that interfered with telegraph poles and wires. The crew had decided after their earlier inspection that this tool would be ideal to cut the metal straps that restrained the solar wing. However, when Kerwin tried to use it he found that it was impossible to place the cutting jaws precisely where he wanted them, partly owing to the length of the handles, but mainly because he was unable to get the leverage he needed for his own body in the weightless conditions. After many exhausting attempts, he noticed an attachment point on the hull and by connecting his dual tethers to this point, and one other, he discovered that he could "stand" on the hull with the tethers strained against him. This gave him the leverage and positioning that he needed, and he was able to snap first one of the restraining straps, and then the other. Almost unbelievably, the solar wing refused to deploy. Both men looked on in exasperation, until it was realized that the hinges were probably frozen and holding the wing in place. Kerwin decided to venture out into the middle of the wing and push against a rope that was tied to it, and eventually the hinges were freed and the wing began to deploy. Conrad, meanwhile, had been shot from the wing like an arrow; but his umbilical line caught him and he returned to the station hand over hand in time to see the wing fully deploy. Conrad and Kerwin re-entered the station whilst delighted ground controllers confirmed that the wing was now fully deployed and generating electricity, Skylab was saved.

Conrad and his crew could now settle into more of a standard routine, more like the one originally envisaged. They immediately discovered that Skylab was big and roomy, much larger than any spacecraft they had previously experienced. To give some idea of its size, the interior usable volume of Skylab was about 361 m^3, which is a fairly meaningless number; by comparison an average semi-detached three-bed-room house has a volume of about 270 m^3. That made Skylab pretty big, but bear in mind that in your three bedroom house on Earth, in normal gravity, you only get to use the floor space of that 270 m^3, any space above your head is essentially wasted. In orbit, in zero-g, *all* of that space is habitable whether its floor, ceiling, or wall. The early Salyut stations had little more than 100 m^3 of space so you can see that Skylab was large for its time, and in fact its internal size would not be surpassed until the Mir space station had been fully constructed twenty-five years later.

The hydrogen tank that the crew now lived in was split into two decks, if you imagine Skylab standing upright as it was on the launch pad with the workshop at the bottom, and the docking adapter and telescope mount at the top. The very first thing we see working from the bottom of our stack, is the original oxygen tank of the Saturn rocket stage, this tank has been basically left alone, and was used to store all of the crew's rubbish. The crew put the rubbish into the tank via an airlock connector which ran between the oxygen tank and the much larger hydrogen tank. The "bot-tom" floor of the hydrogen tank contained the crew's individual sleeping quarters, the ward room, the bathroom, an experimental rotating chair, and the airlock for the rubbish tank, as well as a shower, a first for any manned spacecraft. Each crewman had his own sleeping compartment, with a sleeping bag hung on one wall, and storage space for personal items. Pete Conrad found that he did not like the way his sleeping bag was hung because the airflow went up his nose, so he turned the sleeping bag around; of course it's all the same in zero gravity. The wardroom contained a table that all three crewmen could assemble around with a separate area for each of them; this allowed them to heat their food with a kind of tray to eat from. In the center of the table there was a water dispenser, both for drinking directly from, or for re-hydrating their food packs. The table also included a kind of bar stool arrangement for each man, but they found these very awkward to use as it meant that they had to conscientiously bend over the whole time, and their abdominal muscles quickly became tired. The shower, which many might think would be a very welcome addition to any spacecraft that you are going to spend a significant amount of time aboard, proved to be not as useful as hoped. The shower compartment was not a permanent glass structure that you might expect on Earth, but a collapsible enclosure to aid cleaning. In the absence of gravity the water had to be pressurized for it to "flow" from the shower head, and the water had then to be collected by means of a suction head much like a vacuum cleaner that was used to suck the water from the interior of the shower, and the astronaut. The crews found that whilst it was a pleasant experience to have this facility, it took a great deal of time to set-up, use, and clean up after, and they therefore used it less often than they otherwise might have. The bathroom was not quite such a chore to use, but the three crews did all find it a little odd that the designers had chosen to place the toilet on the wall, which meant that the crewman ended up facing the floor. In all other respects that system worked well, which was just as well, as the alternative meant reverting to the Apollo plastic bag method!

The main reason for the crew's presence on board, of course, was to carry out scientific experiments. A great many of these were carried out on the crew themselves, to study the effects of long-term weightlessness on the human body. One of the other important roles of Skylab was to study the Sun. An entire suite of equipment had been designed for this purpose, and the crew trained extensively in its use. Once the power problems were solved, the crew were able to carry out their full schedule of Sun observations using the ATM (Apollo Telescope Mount).

An important milestone was achieved on 17 July when Conrad's crew surpassed the 23-days-in-space mark set by the Soyuz 11 crew on board Salyut 1 in 1971. They spent their final week finishing the current experiments, stowing results for return to Earth, and getting the station ready to be unmanned for a period of time before the arrival of the next crew. Once the crew had separated from Skylab, another fly-around was carried out to photograph the condition of the station, then they fired the SPS engine to initiate the return home. The crew had completed 28 days in space, and Conrad was now the new spaceflight record-holder with over 1,179 hours in space. Years later, when asked, he would say that Skylab 2 was the mission that he was most proud of, and that when he thought about space, he always thought of Skylab and all of that room. Most people he met assumed that his mission to the moon would have been the highlight of his career, but as far as he was concerned Apollo 12 had gone by the numbers, and had been relatively routine; he would not trade it for the world, but it really had not been that exciting. Skylab was different; he and his crew had faced unknown problems, and surmounted them, and they had left the station able to continue the mission for which it had been launched, as well as achieving nearly all of the mission's scientific objectives.

Skylab's mission continued after the departure of the first crew. The ATM had been designed to be controllable from the ground, and therefore solar observations

Skylab 3 crew in front of Pad 39B

continued. Unfortunately, a primary gyroscope used to control the orientation of the station failed, and observations were stopped until the next crew could arrive. It was decided to bring forward the launch of Skylab 3 so that they could replace the failed gyro, and also install an improved sunshield, as controllers feared that the temporary solution deployed by Conrad's crew was deteriorating faster than expected.

The Skylab 3 crew consisted, as planned, of commander Alan Bean, pilot Jack Lousma, and science-pilot Owen Garriott, and their command module was almost as packed with additional items as the first crew's had been, partly because the intention was to increase the mission duration by three days, to the originally planned 56 days. The improved sunshade was one thing, but they also carried extra film canisters, extra food, various spare parts, including a replacement set of gyros. Launch was set for 28 July 1973, and the countdown proceeded smoothly. Only Bean had flown previously; Garriott and Lousma were rookies. Lousma fell asleep whilst waiting for lift-off. As he would later recall, "Just about thirty seconds before launch, you reach over to your buddies, shake their hands and wish them good luck, because their luck is going to be the same as yours!"

Skylab 3 was launched flawlessly, and had no trouble docking with Skylab. After many checks, the crew entered the workshop to mark the first time that a space station had been reoccupied by a different crew. However, the mission had not been without some complications at this early stage. Lousma had started to suffer from some "stomach awareness", or Space Adaption Syndrome as we now call it, shortly after reaching orbit, and later as they entered the station Bean and Garriott had also begun to suffer too. Bean had, of course, flown on Apollo 12 with no problems at all, and it caused some surprise in Mission Control when he reported feeling ill. The crew did their best to carry on with their duties, but inevitably fell behind schedule. The net effect was that Mission Control tried to give the crew additional rest time in an effort to speed their recovery, and also postponed the first planned EVA by 24 hours. Over the next couple of days, the crew slowly began to feel better and began to catch up on the schedule; however, the entire episode caused concern for mission planners, especially with the next Skylab crew—all rookies—scheduled for a longer mission.

The problems did not end there unfortunately. It had also been noticed early on that one of the thruster quads on the Apollo service module had sprung a leak, and eventually it was deactivated. The spacecraft was able to fly perfectly well with the three remaining quads. However, several days later a second quad also started to leak and had to be shut down. This still did not represent any immediate danger for the crew, as Apollo was quite capable of flying on two, or even one thruster quad, but it did cause concern that eventually all four quads might be rendered useless. NASA's contingency planning came into its own at this point; a rescue mission had been planned for all three missions to Skylab, and it was this option that saved the mission. If there had been no possibility of a rescue mission, the Skylab 3 crew would have packed up and come home as soon as possible, whilst the two remaining quads were still operational. But the possibility of flying a rescue command module meant that both the crew and Mission Control could afford to wait and see. In the meantime, the rescue crew of Vance Brand and Don Lind rehearsed in the simulators and their modified command module was readied for flight. The engineers on the ground were able to determine that the leaks in the two thrusters were unrelated, and that there was nothing to suggest a systematic fault. The rescue crew were stood down, although they did spend time simulating the Skylab 3 return with only two working thrusters. Lind would later remark that he had effectively talked himself out of his first flight by showing that the Skylab 3 crew could return safely without the need for a rescue flight.

After these dramas, life settled into a gentler routine for the Skylab crew. There were a few equipment malfunctions that had to be attended to, but on the whole the rest of mission was quiet. Garriott and Lousma installed the improved sunshield during an EVA 10 days into the mission. The same pair also retrieved film cassettes from the ATM later in the mission, and later still Bean and Garriott retrieved more film cassettes and also retrieved a sample of the new parasol to determine its condition after a month's exposure. When the time came for the crew to leave the station, they had more than completed their objectives, and after the initial problems with space sickness had subsided, had consistently been ahead of the flight plan, always asking for more work, and by the end of the mission they had in fact achieved over 150% of their targeted work. Whilst this was a fantastic achievement, it would not bode well for the crew that was to succeed them.

Skylab 3 rescue crew

The crew for the third and final Skylab mission broke from Deke Slayton's usual rules of crew selection; they were all rookies. Their mission had changed somewhat, too. A comet had been discovered that would approach the Sun toward the end of 1973, and the launch of the third crew was delayed from its original October launch date until November so that they could carry out observations using Skylab's ATM and other instruments. The booster for the last Skylab mission had been sitting on the pad for some time, as it had originally been rolled out to serve as booster for the Skylab rescue mission; when this mission was stood down, the booster became the Skylab 4 launch vehicle. However, just five days from launch a routine inspection crew discovered cracks on the stabilizing fins of the first stage. Perhaps this was not surprising, as this stage had been manufactured over seven years earlier, but clearly it

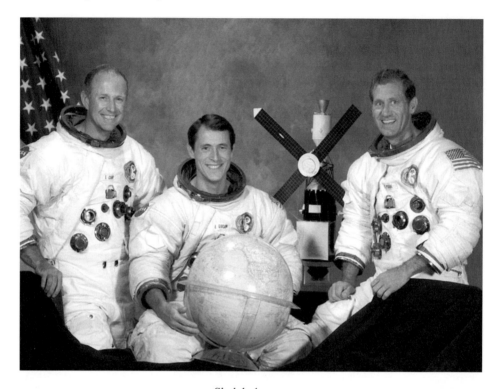

Skylab 4 crew

could not be launched in this condition. It was decided to replace the fins on the pad, which would take about a week. The crew faced a tight squeeze in their Apollo Command Module due to it being packed with additional items for the long mission ahead, most of it food to allow the length of the mission to be extended from the planned 70 days to 84 days if all else was well. The launch itself was routine and seven hours later the crew sighted Skylab and prepared to dock—which they had some difficulty with initially, but managed at the third attempt.

With the experiences of Al Bean's crew very much in mind, Mission Control had ordered the astronauts to take more precautions against space sickness in order to prevent disruption to the early mission flight plan, and they took anti-sickness pills as soon as they reached orbit. It was also decided that the crew would have a sleep period before entering the station for the first time. Unfortunately, it was swiftly proved that this approach did not help, as Bill Pogue was overcome with nausea almost as soon as the rest period began, and relieved himself of his last meal. The crew made the first mistake of the mission when they decided not to mention Pogue's symptoms to Mission Control. Confident that he would feel better before they entered the lab for the first time they simply explained that he had not felt hungry and had left most his last meal uneaten. This plan might have worked if it were not for the on-board automatic taping system which recorded the entire conversation and relayed it later to the ground, and most importantly to Chief Astronaut Alan Shepard. As a result Shepard talked directly to the crew commander, Gerry Carr, and voiced his opinion on what he called "a fairly serious error in judgement". Carr realized the error of his ways and put his hands up and agreed that "it was a dumb decision".

Despite the best efforts of the mission planners, Pogue's sickness would impact the early activation of the station by limiting his participation with the rest of the crew. In fact, the planners seemed to assume that this crew could pick up at the same pace as Bean's had left off, which ignored the fact that it took Bean's crew several days to get near that pace of work. The planners also seemed to assume that procedures in space took the same amount of time as taken during training on the ground, and as hard as the crew tried to keep pace, they simply could not, and fell further behind the timeline set by the ground controllers. Even worse, the planners on the ground did not seem to realize that they were making things worse; they even added extra tasks to the crew's day, causing them to fall even further behind, and consequently start to believe that they were not doing a good enough job. On the seventh day of the mission, Pogue and Ed Gibson carried out a planned EVA to replace film cartridges successfully, but even then the tired crew left some stowing away tasks until the next day. All in all, the first three weeks or so were very difficult. But things began to improve as the crew realized how to make things better, and better communicate those thoughts to the controllers on the ground. This was about the same period of time that Al Bean's crew had taken to reach their peak efficiency, but this fact was apparently forgotten by the mission planners, who seemed to assume that the new crew could immediately start where the previous crew had left off. The crew desperately tried to remain on the timeline, and explain the problems to those on the ground, but their pleas went unheeded. The mission planners, for their part, always felt that the crew were about to reach their best performance level, and were therefore reluctant to reduce the workloads. After all, this was the last chance for these scientific experiments to be flown and NASA wanted to take advantage of every waking moment. It all came to a head after the crew had been in orbit for about six weeks. During a call with the crew's boss, Deke Slayton, all of the problems were voiced and discussed, the ground were persuaded to ease off on the workload, and also leave some of the scheduling to the crew rather than providing a daily minute-by-minute task list. This meant that the crew felt more in charge of their activity, and were able to follow a more "normal" day. The rest of the mission proceeded at a similar pace to the previous ones, and by the end of January 1974 the crew were making preparations to return home. The orbit of Skylab was raised slightly with a firing of RCS jets on the Apollo service module, in the hope that this might allow Skylab to survive for longer, and perhaps be visited again before its expected orbital decay in 1981 or 1982. The Skylab 4 crew landed about 5 hours after undocking having spent a total of 84 days and 1 hour in orbit.

The possibility of a Skylab revisit and re-boost mission would now be left to the space shuttle, which at this stage did not exist, so a choice had to be made between trying to preserve the station for some future visit by Apollo CSM or the space shuttle, or a mission to send a crew in Apollo to carry out a controlled re-entry burn to send Skylab to its destruction. There were some risks attached to the latter, as it involved the docked CSM firing its service module engine until Skylab had almost reached entry interface, which meant that a prompt undocking was a very important action; if the docking latches failed in some way the crew would follow Skylab to destruction! In part due to these risks, it had been decided to boost Skylab to a higher orbit before the final crew left, effectively deferring the decision until the early 1980s. Once the space shuttle program was underway, it was tentatively planned that during its third flight the shuttle would rendezvous with Skylab and attach a booster rocket to the docking port, at which time it would be decided whether to boost the station to a higher orbit once more, or send it to the bottom of the Pacific Ocean. Ironically, Jack Lousma of Skylab 3 was assigned to pilot the shuttle's third mission, and revisit his old home. In the end, two factors decided Skylab's fate. The first was the protracted development of the shuttle, it became clear over time that the shuttle would simply not be ready in time to save the orbiting station, especially as it's orbit was deteriorating faster than expected owing to increased solar activity inflating the upper atmosphere and causing increased drag. Skylab would have to be left to make an uncontrolled re-entry sometime in 1979, and it seemed every nation in the world was worried that it would fall on them. Shortly before its crash to Earth, it was determined that Australia was the most likely target, and at least 25 tons of various parts of the station were predicted to survive the re-entry process. In the event, several parts did survive, and a young Australian claimed the $10,000 prize that a U.S. newspaper had offered as reward for any genuine Skylab parts. The largest items found were a door from one of the film vaults, and some oxygen and nitrogen tanks, and these along with various museum pieces like the back-up Skylab are all that remain of the United States' first space station.

Was Skylab a success? The answer is both yes and no. Yes, because NASA successfully carried out a great deal of science during the three manned periods, and even during the unmanned intervals as well. For an agency that had no real experience of carrying out scientific experiments, other than those on the surface of the moon, and none at all over long periods of time, it was a very successful project. Detailed photography and data about the Sun was collected—enough to keep researchers busy for some years, human medical experiments, materials processing, and more besides, were all carried out with precision and accuracy by the various crews. On the other hand, Skylab was not a success, because the mission planners in particular seemed unable to learn from the experiences of previous crews. The work schedule for all of the crews was always unrealistic. It was an easy mistake to make on your first space station project; the Soviets had experienced similar problems after all with the early Salyut mission. Amazingly, NASA would be doomed to repeat these mistakes in years to come on board Mir and the International Space Station.

4

1973–1974: Salyut 2 and Salyut 3—limited success

Salyut 1 had not been a failure; the death of the returning crew was tragic, but the space station itself had been blameless. The Soviet space organization was keen to launch another as soon as possible. Of course they had to wait until the Soyuz ferry vehicle was ready to resume flight. Arguments raged over the best way to achieve a safer Soyuz design, Chief Designer Vasily Mishin argued that simply adding space-suits to the capsule was not the right answer, it limited the crew to two and seriously reduced the cargo the ferry could carry to and from orbit in addition to the crew. However, he was overruled by the Secretary of the Central Committee, Dmitri Ustinov, who was absolutely determined that no cosmonaut would be launched into orbit without a spacesuit ever again. Mishin continued to argue that reliability of the systems was the best method of ensuring the safety of the crew, but he was told in no uncertain terms that Ustinov's word was final, and that pressure suits were to be installed. This required a new version of the Soyuz spacecraft that would only have an orbital lifetime of two days. This was necessary as the solar arrays of the previous version had to be removed to save weight, and the on-board batteries would only last for that limited time.

After the "civilian" first station, it had been decided to introduce the first pure Almaz design, designated OPS-1, albeit using the Soyuz craft as the ferry instead of the TKS. The Almaz design also differed from the first Salyut in that its docking port was at the rear of the station. OPS-1 made it to Baikonur in the midst of the harsh winter of January 1973, and during the next 90 days military testers and civilian specialists prepared it for launch. The OPS-1 blasted off into orbit on 3 April 1973. Since the authorities did not want to disclose the existence of two space station projects in the USSR, and particularly, to reveal the development of the military Almaz, the OPS-1 was announced as Salyut 2 upon reaching orbit. It was given the Salyut name to disguise its military configuration, but it was different to the space station that had preceded it. It was several meters shorter, but weighed about a ton and a half more. In the center of its living compartment a huge camera was installed

Proton launcher with Salyut 2

in the "floor" and the station had a much higher level of automation to reflect the reduction in the Soyuz crew.

Unfortunately this new station did not last long enough for any crew to board it, and perhaps this was just as well, because 13 days after launch an electrical fire in the propulsion unit spread to the main compartment, explosively decompressing the station and sending it spinning out of control until it broke up. The official investigation concluded that as a result of a faulty welding, one of the lines in the station's propulsion system had burst during an engine firing and the plume of flame had burned through a pressurized hull. However, future findings were to cast doubt on this theory. Careful analysis of fragments detected in orbit, showed that three days after the launch of the OPS-1 the upper stage of the Proton rocket that had delivered the station apparently exploded as a result of pressure changes in its tanks resulting from overheating. The stage carried about one tonne of unspent propellant, and the explosion created a cloud of debris in the proximity of the station. The speed of some debris differed from that of OPS-1 by as much as $300\,\mathrm{m\,s^{-1}}$. Eight days later, a piece of this orbital junk apparently hit the station. However, despite all this secrecy and the attempt to cover up the military nature of the station, western observers almost instantly managed to discern the military nature of the new spacecraft. An article appeared in the September 1973 issue of *Aviation Week*, which read, "Soviet penchant for secrecy within its own space program has lead to a widespread, but erroneous, belief that a Salyut spacecraft failed while in orbit. The spacecraft, which the Soviet press and information agencies called a Salyut, was launched Apr. 3 and apparently suffered a catastrophic failure on Apr. 14. However, the spacecraft transmitted on a different frequency than previous Salyuts and now is believed to have been a different spacecraft. The reports initially issued by the Soviets apparently were incorrect because of an attempt to keep secret the actual nature of the spacecraft. Telemetry transmissions from the spacecraft were similar to those monitored earlier from Soviet reconnaissance satellites." Although the Almaz name was not known in the West for many more years, these stations had become identified in the West as "military Salyuts".

Undeterred, the Soviets launched another station only a month later. This was not an Almaz, but the third station in the DOS series. DOS-3 carried several improvements over the earlier configuration. It had improved solar arrays in an effort to double the overall lifetime of the station from the 90 days of the two previous stations. In most other respects, however, the station was basically the same, although more thought had been given to automating systems to accommodate the reduced crew of two. Unfortunately, due to errors in its flight control system, and while out of the range of ground control, the station fired its orbit-correction engines until it ran its tanks dry, and a week later re-entered the Earth's atmosphere. Since the station had reached orbit, and therefore been tracked by western ground stations, the Soviets had to acknowledge its existence but, in a effort not to give anything away, designated it Cosmos 557 as a form of disguise.

The Soviets finally enjoyed some success with the launch of Salyut 3, a station of the Almaz design, in June 1974. The OPS-2 space station really was a reconnaissance platform, for it housed a massive 6-meter camera in its main compartment and had a capsule for the high-resolution film to be returned to Earth independently of the crew. In all 14 cameras were to be used on board. The other notable feature of this station was the modified aircraft machine gun that was mounted near the front port for station defence! In order to point it at the target the crew had to change the attitude of the entire station.

When the Salyut 3 station was launched a small group of teachers and school children in Northamptonshire, England, were glued to their radio receivers. The group would later become known simply as the Kettering Group, but for now physics teacher Geoffrey Perry and some of the pupils from Kettering Grammar/Boys School

Salyut 3 Proton launch vehicle

had no particular name. They had for some years been following the satellite launches from both the U.S.A. and the Soviet Union by means of radio receivers within the school. He and his team of teachers and pupils had amassed a great deal of knowledge, particularly about Soviet satellites, since the launch of Sputnik 2 in 1957. Mr. Perry correctly identified launches from a site other than Baikonur in 1966, which would be later known as Plesetsk, and was also the first to record evidence of the first unmanned Soyuz test after the tragedy of Soyuz 1. Now with the launch of Salyut 3, they hoped to positively identify the station for themselves. But when they listened to the same channels that they had listened to previously for manned Soyuz

Salyut 3 on the ground before launch

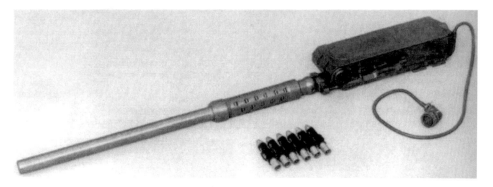

Salyut 3 gun

missions and the Salyut 1 station, they could hear nothing. The same thing had occurred the previous year with Salyut 2, and they wondered why this could be. It was later determined, with the help of other radio amateurs, that the station was using a frequency only usually used by military reconnaissance satellites. Although it was immediately apparent that this change of frequency meant that the station was military in nature, it was clear that it was being operated differently than the previous Salyut station. They hoped to confirm their hypothesis when the inevitable Soyuz spacecraft was launched to dock with the new station.

The crew of Soyuz 14 were Pavel Popovich and Yuri Artyukhin, both military officers, who were launched on 3 July 1974 and docked later that same day.

Crew of Soyuz 14

According to Popovich, the on-board automated rendezvous system delivered the Soyuz spacecraft only 600 m from the station and from a distance of 100 m the crew switched to manual control. Popovich remembers taking off his spacesuit gloves, (unpressurizing his suit, as a result) in order to make it easier to control the craft. The Kettering Group were able to identify for themselves that the Soyuz was only manned by two cosmonauts, and determined the identity of its commander, and they made a press announcement to the world through Reuters before the Soviet Union had even officially announced the launch. By comparison the U.S. CIA had no such detailed information; in a National Intelligence Estimate dated December 1973 they seemed only to be aware of the civilian Salyut program, and gave no indication that they had any information regarding the military Almaz program.

Popovich and Artyukhin entered the OPS-2, having docked at the rear, on 4 July 1974 and spent 15 days on board. According to official sources, the "remote-sensing equipment" was activated on 9 July, followed by several days of photography of the "Earth surface". Central Asia was among officially disclosed targets of the station's cameras. Western sources also say a set of targets laid out near Tyuratam was photographed to test the capabilities of the surveillance hardware. Several times during the mission, an on-board alarm system woke up the crew; however, it proved to be false. During the flight, the cosmonauts reportedly checked the systems on board, adjusted the temperature inside the station, moved some ventilators and completed other housekeeping chores. They also reloaded the station's on-board cameras and placed exposed film into the space station's return-to-Earth capsule.

The second crew consisted of commander Genadi Sarafanov and flight engineer Lev Demin, and they were launched on board the Soyuz 15 spacecraft on 26 August 1974. However, problems with the rendezvous system on board the Soyuz during the approach to the station forced officials to cancel the docking attempt. The spacecraft returned to Earth after a two-day flight, the limit of the Soyuz's orbital endurance, and was forced to land under night-time conditions. Typically for the period, official sources reported only that the Soyuz 15 crew "tested various rendezvous modes during its mission".

Two decades later, the official history of RKK Energia revealed that when Soyuz 15 reached a distance of 300 m from the station, the Igla ("Needle") rendezvous system, failed to switch to the final-approach mode and instead started implementing a sequence that would normally be executed at a range of 3 km from the station. On commands from the Igla, the Soyuz fired its engines, accelerating itself in the direction of the station. The relative speed of Soyuz 15 to the OPS-2 reached $72 \, km \, h^{-1}$, zooming by the station at a distance of 40 m. As the crew failed to realize the problem (and to shut down the Igla), the rendezvous system attempted to re-acquire radio-contact with the target and sent Soyuz 15 to the station twice more each time narrowly avoiding a collision. By the time ground control commanded the deactivation of the Igla, the crew only had enough propellant for the descent back to Earth.

Due to lengthy modifications in the wake of Soyuz 15's rendezvous problems, no further expeditions to Salyut 3 could be staged. The return-to-Earth capsule was jettisoned from the OPS-2 on 23 September 1974 and successfully recovered on Earth, and the station was de-orbited on 24 January 1975 over the Pacific Ocean.

According to official Soviet sources, the seven-month flight of Salyut 3 exceeded more than twice the originally planned flight duration. Soviet publications also disclosed that Salyut 3 was the first space station to maintain constant orientation relative to the Earth's surface. To achieve that, as many as 500,000 firings of the attitude control thrusters had been performed. This fact also hinted to Western observers that Salyut 3 had perhaps carried out a reconnaissance mission.

Years later it was revealed that shortly before de-orbiting OPS-2, ground controllers commanded the "self-defence" gun on board the station to fire. According to Igor Afanasiev, an expert on the history of space technology, firings were conducted in the direction opposite to the station's velocity vector, in order to shorten the "orbital life" of the cannon's shells. A total of three firings were conducted.

5

1974–1977: Salyut 4, 5, and ASTP

Salyut 4 differed from Salyut 1, the previous successful DOS design, by having three sets of solar arrays, just as the doomed Cosmos 557 had. It also included some additions for crew comfort, including a table for the crew to eat at which supplied hot and cold water for rehydrating their food packs. The navigation system for the station was now semi-automatic, to allow the crew more time for experimentation. It was launched on 26 December 1974, and was followed on 11 January 1975 by Soyuz 17 with a crew of two, Alexei Gubarev and Georgi Grechko, both making their first space flights.

Their docking was achieved effortlessly, and they soon settled into the mission, working for six days a week, with a day off to spend largely as they wished. Their enthusiasm for their work was such that they worked longer hours than anticipated, and also ate more than planned, which had to be controlled as there were only so many supplies on board. Eventually they were told to slow down and take more time off, which they reluctantly did. The Soviets were still working to discover the best compromise between work and rest for the cosmonauts' working week. The crew returned to Earth on the 7 February after 30 days in orbit, a new Soviet record of endurance, and were found to be in good physical and mental shape. However, it was decided that the exercise regime for future crews would be stepped up slightly, especially in the later stages of the mission to ensure that they were in the best condition for re-entry and adaptation to Earth's gravity.

The launch of Soyuz 18 on 5 April was rather more dramatic, and once again the Soviets failed to get a mission to a space station. A fault with the separation of the main booster stage caused the abort tower to be used for the first time in manned spaceflight, causing the crew, Vasili Lazarev and Oleg Makarov a very uncomfortable 15-g ride before the capsule landed in snow. The next launch attempt on 24 May was also called Soyuz 18, and the previous failed flight simply referred to as "the 5th April anomaly". This crew, Pyotr Klimuk and Vitali Sevastyanov, reached orbit successfully, and docked with Salyut 4. Their task was essentially to carry on the

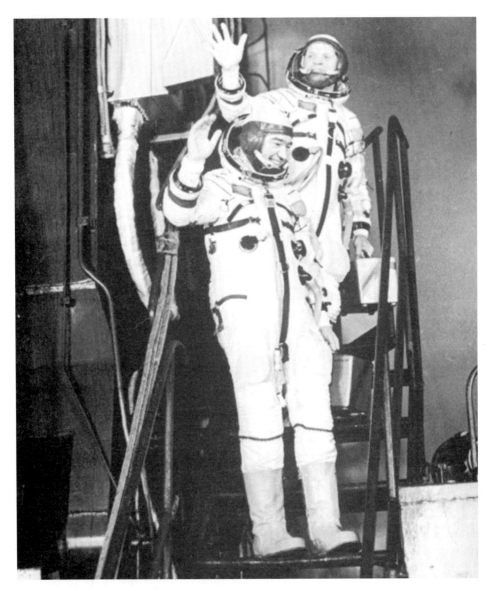

Soyuz 17 crew shortly before lift-off

work started by the previous crew, but this time the Soviets tried to schedule the workload more logically. They would work for several days on one type of experiment before moving on to the next; a significant step forward. Of course, there was also maintenance work on the station to be carried out between scientific experiments, but again experience from previous missions was paying off, as items such as filters, pumps etc., had been made much easier to replace and service than previous designs, cutting down on time and frustration.

A new mission was launched on 15 July 1975. Although this mission did not involve Salyut 4, it did signify the first co-operation between the Soviet Union and the United States of America: ASTP, or Apollo–Soyuz Test Project. Discussions for such a mission had been taking place for several years, as mentioned in an earlier chapter, but the options were soon narrowed down to one: the docking of an Apollo Command and Service Module with a Soyuz spacecraft. On 24 May 1972, U.S. President Richard Nixon, and Soviet Chairman Alexei Kosygin, signed the agreement that would make ASTP a reality. It was part of a much larger agreement that covered all manner of scientific co-operation, including space flight. Despite the air of co-operation, there remained "discussions" about various aspects of the mission. Which spacecraft would launch first? NASA assumed that Apollo with its longer mission duration would be the first to launch. That way if Soyuz was delayed for any reason, Apollo could simply wait until it arrived. The Soviets disagreed, stating that they would launch first, and wait for Apollo; if Apollo were delayed, they would launch a second Soyuz if necessary. This came as something of a surprise to NASA mission planners, as they had not previously heard anything about a second Soyuz being prepared for this mission. The actual docking posed even more problems, both technical and political. What form would the docking mechanism take? Both nations so far had used a male and female docking mechanism; which nation would take which role? The Soviets were rather more chauvinistic in this area, not wishing to take the "lesser" role of the female as they saw it. It was finally agreed that the docking system would be an androgynous one that equalized the two nations, but this also presented another problem. Which spacecraft would be the active (moving) ship, and which would wait (stationary) for the docking? The Apollo was clearly the more maneuvrable spacecraft, and so the Soyuz would have to wait to be docked with by the Apollo. Unfortunately, the problems did not end there. The atmospheres of the two spacecraft were very different. Apollo's atmosphere consisted of 100% oxygen at a pressure of 0.34 atmosphere, whilst the Soyuz was an oxygen/nitrogen mix at 1.0 atmosphere. Clearly, it would be possible to simply float from one spacecraft to the other without suffering from the bends, so the docking mechanism would also have to double up as an airlock. With modifications to both spacecraft to allow the lowering of cabin pressure to make the transfer between the two craft quicker achieved, the major technical problems had been overcome. The first of the two crews were announced in 1973; the U.S. crew would consist of commander Tom Stafford who had previously commanded Apollo 10 and Gemini 9, and flown as pilot on Gemini 6. The Command Module Pilot would be Vance Brand, who would be making his first space flight after backing up the last two Skylab crews. The third crewmember had been waiting for a flight for a long time, after years of selecting other astronauts for their missions Deke Slayton, a member of the original Mercury astronaut group, would finally make it into orbit. The Soviets followed with their own crew announcement a few months later. This was a first, as crews for Soviet space flights had never been announced in advance before. The commander of Soyuz 19 would be Alexei Leonov, who we know all about from his travails during the crew selection for Soyuz 11. His flight engineer would be Valeri Kubasov, who had been removed from the Soyuz 11 crew for medical reasons, but was long since recovered. He had flown

U.S. ASTP crew

previously on Soyuz 5 and had spent time on Salyut 6; in fact he had more flight experience than his commander.

With the crews announced, training for the flight could now begin. Learning each other's language proved to be the most difficult task for both crews; something that would not change much over the coming years. The problem did not end with the crews of the spacecraft, the ground controllers and technical experts also needed to get up to speed on their counterparts' language, a task that is particularly difficult where technical jargon is concerned.

Apollo–Soyuz finally got underway when Soyuz 19 was launched on 15 July 1975. With Soyuz safely established in orbit, Apollo was launched to give chase. Almost two days later, the two spacecraft docked without difficulty. The two crews spent 47 hours docked together, with members of each crew visiting the other's spacecraft. Mission rules dictated that neither vehicle would be left unmanned at any time. After the docked phase of the mission Soyuz 19 returned to Earth almost immediately even though this Soyuz was equipped with solar panels like previous versions, and could stay in space for longer than its space station specific counterparts. Apollo stayed in orbit for a further three days to conduct experiments that

Soyuz 19 crew

Combined U.S. prime, back-up, and support crews

would have to last NASA for a while, this being the last U.S. manned mission until the space shuttle was ready to fly, at this time expected in 1979.

It has been suggested that ASTP was little more than a political show, but this opinion sells the program short. In truth the idea may have been born out of political needs by both the U.S. and the Soviet Union, but that fails to take into account that many of the people that worked on this mission from crews to support staff and technical designers from both countries would later work together again on Shuttle–Mir, and ultimately the International Space Station (ISS). Relationships that were forged during ASTP would endure to smooth new relationships in the 1990s. It has been said that neither side learned very much; the U.S. engineers say that most Soviet equipment was Gemini era to them. But both countries did learn that it was possible to work together, and in the longer run that was sure to be worth something.

Meanwhile the crew of Salyut 4 continued their mission for a few more days before they too returned to Earth on 26 July, after 63 days in orbit. Before their departure they fired the engines of Soyuz 18 to raise Salyut 4's orbit. This was the first time that such a maneuvre had taken place, previous orbit-boosts had been undertaken using the stations own propulsion. Upon their return the crew walked from their capsule to the medical tent, obviously in better physical condition than previous crews had been. Their mission had paved the way for longer duration flights, possibly involving rotating crews on a new space station. But Salyut 4's mission was not over. Soyuz 20 was launched on 17 November 1975, and unusually for such a designation it carried no crew. It followed a different flight profile than usual, docking with Salyut 4 after two days rather than the one day that had been flown by manned up until now. Once docked, Soyuz 20 remained powered down until the end of February 1976. It later became apparent that this mission was a test of the flight profile and duration of an unmanned cargo craft. Once the craft had returned to Earth, examinations of its systems led Soviet engineers to place a 90-day limit on the amount of time that a Soyuz could safely spend in space.

Salyut 4 had been a great step forward for the Soviet space station program, after the difficulties of its predecessors, and had proved the procedures and technology that would be needed for the next generation of stations.

The next space station was Salyut 5, and it was launched on 22 June 1976. It quickly became apparent from the telemetry that this was another Almaz reconnaissance platform, identical to Salyut 3, but without the machine gun. An all-military crew of Boris Volynov and Vitali Zholobov was launched on Soyuz 21 two weeks later. The crew carried out some scientific experiments, but their primary mission seemed to involve observations of a military exercise that was underway in Siberia. They seemed set for a fairly long-duration mission, and indeed Soviet radio had reported on 19 August that solar radiation levels were such that the crew would be able to carry out a "prolonged flight", but five days later the same radio station reported that the crew were in the process of returning home. When they clambered out of the capsule after a night landing, it became evident that they were suffering from the effects of their mission, most likely because they had not started their pre-return exercise regime. All evidence seemed to point to the fact that the crew had returned much earlier than planned; but why? There were several suggestions, but the most likely seemed to be that the station's atmosphere had somehow become contaminated causing the crew to abandon ship. One speculation in the West was that one or both crewmembers had suffered mental or physical problems that forced their return. It was also suggested that Zholobov in particular had suffered debilitating homesickness, and as a result had not followed his exercise regime. It has since come to light that perhaps the two crewmembers did not get along, and perhaps their hostility got to the point that returning them to Earth was the only option before physical harm was caused. Whatever the reason, Salyut 5's first manned mission had been abandoned early, and a crew needed to return as soon as possible to carry on the work. The next flight, that of Soyuz 22, made use of the back-up vehicle from the ASTP mission, and was not a flight to Salyut 5. Soyuz 22 was a week-long flight that concentrated on Earth photography using a special East German built camera. The fact that the next crew to visit Salyut 5 was launched within 2 months of the landing of the previous one, suggested that not too much could have been wrong with the station. The crew of Vyacheslav Zudov and Valeri Rozhdestvensky were to check on the condition of the station, and carry on with the experiments that were still left on board. However, all was not to go smoothly. The mission of Soyuz 23 on 14 October proved to be dramatic. The automatic rendezvous system malfunctioned almost as soon as it reached orbit, and for some reason not made clear by Soviet officials the crew did not attempt a manual approach and docking, but waited for the first opportunity to land. When the landing attempt came, things began to go wrong. After re-entry, the capsule descended on its parachute in the darkness, but was blown off course by a blizzard and it landed in Lake Tengiz. This was not in itself a problem as the Soyuz had been designed to land in water if necessary. Although the capsule landed in relatively shallow water, it was at least 5 miles from the nearest shore, and the water was freezing. Recovery by boat was therefore impossible, and helicopters could not locate the spacecraft as thick fog engulfed the area. The crew were forced to spend a very cold night in the capsule, which by now had no power reserves, and therefore no heating. At first light, the fog had cleared sufficiently to allow helicopters to attach lines to drag it to shore and end the exhausted crew's ordeal. As it was, no real damage was done, but it is worth reflecting on what might have happened to a more weakened crew that had perhaps spent months in orbit, or been forced to come home early ...

Viktor Gorbatko and Yuri Galzgov were launched on 7 February 1977 on Soyuz 24 for what would prove to be a short mission. Again when an automatic docking was attempted the system failed, but on this occasion a manual docking was attempted and achieved. The crew entered the station wearing breathing apparatus, but it did not take long to determine that the air was clear. Whether this procedure was really necessary, only the Soviet space officials know. However, the crew did take the opportunity to test a new procedure to clear the station's atmosphere of any potential contaminants. They had brought some equipment with them that would allow the existing atmosphere to be vented and replaced with fresh supplies of air stored on board. Basically, the old air was allowed to leak out of one end of the station whilst at the same time fresh air was pumped in at the other end. Interestingly, the crew remained on board the station whilst this procedure was carried out, rather than retreating the Soyuz as one might expect. Shortly after this test, the crew began to prepare to come home, packing up their experiments as well as those left behind by the Soyuz 21 crew. The combined equipment and experiments were far more than the Soyuz could return by itself, so use was made of Salyut 5's own descent capsule, designed for just such a purpose; it would return to Earth a day after the crew. The station had been left in a good state of repair, and Anatoli Berezovoi and Mikhail Lisun were already in training to attempt an unprecedented third period of occupancy, but due to the run of ill-luck the Soyuz spacecraft that had been allocated to the Salyut 5 programme had all been used, and there simply was not the budget nor the time to build a new Soyuz during the remaining lifetime of Salyut 5. Soyuz 24, therefore, became the last mission to a military space station. Salyut 5 re-entered the atmosphere on 8 August 1977 when all of its remaining fuel had been depleted, and it was clear that it would host no more missions.

6

Salyut 6: Space station operations defined

Salyut 6 represented the major step forward in space station operations that the Soviets had been planning for some time. Launched on 29 September 1977, it featured a second docking port, as well as an Extravehicular Activity (EVA) hatch. The second docking port was a significant addition because it allowed the station to be resupplied by Progress cargo spacecraft, an essential capability for long-term habitation. It would also allow for the possibility of visiting crews who would dock with the station, stay for about a week, and then return in the older Soyuz leaving the new one for the long-duration crew.

   The concept behind the Progress spacecraft was a simple one, and solved the problem that all previous space stations, including Skylab, had encountered. How do you keep a long-duration crew in orbit, when they are eventually going to run out of supplies of food, clothes, and of course, oxygen? The Soyuz spacecraft could only carry so much cargo in addition to its crew, but if the crew, and all of their life support systems were removed, this released a lot more room for cargo. The result was to become the Progress, essentially a leaned down Soyuz meant only for cargo and fuel, and designed only to make a one-way trip. The heat shield was also removed; it was unnecessary because the idea was that once the resident space station crew had unloaded all of the fresh cargo, they would load the craft with all of their unnecessary equipment, and rubbish, and it would then undock and be remotely commanded to re-enter and burn up. This also made the separate descent module unnecessary, and it was instead used as a fuel tank to allow the Progress to replenish the propellant tanks of the space station. In truth, it would become apparent in later years that the Progress did not solve all of a space station supply problems. It could not return anything to Earth obviously, which meant that crews returning to Earth would continue to need to bring back experiment results with them in the Soyuz, which had a limited return weight. It also turned out that not all of a stations unwanted material could be disposed of in a Progress, and that long-lived space stations would accumulate more and more clutter.

The new ability for crews to make short visits to an occupied station opened up the prospect for the first time of visits by cosmonauts from other countries. In a response to the NASA selection, in 1976, of non-pilot mission specialist astronauts for upcoming space shuttle missions, the Soviets launched the Inter-Kosmos program, with the participation of fraternal communist states, initially Bulgaria, Cuba, Czechoslovakia, East Germany, Hungary, Mongolia, Poland, and Romania in joint space flights with the Soviet Union. In 1979 three non-communist nations were added to this list, France, Vietnam, and India, and all of these flights would be carried out between 1978 and 1983. This agreement led in 1985 to an expansion of the program to all countries, communist or not, organized by GlavKosmos. This led to countries such as Afghanistan, Austria, Japan, and the United Kingdom agreeing to manned space flights with the Soviet Union. All of this, however, lay in the future, for now the first guest cosmonauts were training to fly on board Salyut 6.

Operations did not start well with the launch of Soyuz 25, which unfortunately could not dock with Salyut 6, probably due to a fault in the Soyuz docking system, and had to return to Earth. This resulted in an upheaval in the schedule, as this first crew had been due to occupy the station for about two months, during which time they would receive the first on-orbit visitors, and also oversee the docking of the first Progress cargo vehicle. This failure forced mission planners to attempt a winter launch, which, for safety reasons, they were not generally keen on. However, Soyuz 26 was launched on 10 December 1977 and docked, this time with the rear port, successfully a day later. This mission would set the pattern for all future space station operations to follow. The crew of Yuri Romanenko and Georgi Grechko remained on the station for a record breaking 96 days. In view of the failure of Soyuz 25 to dock, they carried out an EVA to check the front docking port, during which they found nothing out of the ordinary. They received their first visitors when the crew of Soyuz 27, Vladimir Dzhanibekov and Oleg Makarov, docked to the front port and formed the first four-man crew in history. Dzhanibekov and Makarov departed on 16 January 1978 after a six-day visit, taking the older Soyuz 26 home and leaving the rear docking port available for the first Progress cargo spacecraft. The final component of modern space station operations was completed with the launch, on 20 January, of the first Progress cargo spacecraft. This docked at the rear port of Salyut 6 two days later. It was relieved of its cargo and loaded with unneeded equipment, rubbish etc., and then pumped fuel into the Salyut's propulsion tanks. It undocked from the station on 6 February, tested its back-up rendezvous system, and re-entered the Earth's atmosphere two days later. Again, this would become a standard procedure for future space station operations, continuing today with the ISS. This procedure will not change until the ESA's ATV cargo craft starts operations in 2007.

The second crew of visiting cosmonauts on board Soyuz 28 docked with the rear port on 3 March. Alexei Gubarev was accompanied by the first international Inter-Kosmos cosmonaut, Vladimir Remek from Czechoslovakia. The visitors undocked from the rear port after a flight of nearly eight days, this time in the same Soyuz they had launched in, and landed safely. Soyuz 27 undocked from the front port on 16 March with Romanenko and Grechko aboard. Their flight, which had surpassed the record set by the final U.S. Skylab crew of 84 days, had been a tremendous success;

Soyuz 26 crew

they had proved every aspect of space station operations and set the path for all future long-duration expeditions.

The second main expedition to Salyut 6 was undertaken by the Soyuz 29 crew of Vladimir Kovalyonok and Alexandr Ivanchenkov. They launched on 15 June 1978, and docked with the front port the next day. They were to receive two Inter-Kosmos crews, one with a cosmonaut from Poland and the other from East Germany, unload and repack three Progress cargo craft, make an EVA to retrieve material samples from the hull of Salyut 6, and swap the newer Soyuz 31 from the rear port to the front port in order to clear the rear for future Progress dockings. They returned to Earth on 2 November after further extending the duration record to 140 days.

Unfortunately, things would not go quite as smoothly for the third expedition. Vladimir Lyakhov and Valery Ryumin on board Soyuz 32 launched successfully on 25 February 1979, and docked with the front port the next day as per normal. They were expecting to stay for about six and a half months, and apart from working to try

Soyuz 28 crew, first Inter-Kosmos flight

and fix a small leak in one of Salyut's propellant tanks the mission was preceding as planned. Their first visitors were not so lucky. Soyuz 33 as usual contained an international crew with the guest cosmonaut from Bulgaria, but when they got within range of the station the main engine on their Soyuz misfired, and the docking was aborted. The bitterly disappointed crew made a manual re-entry the next day. This had implications for the long-duration crew. What would happen if their ferry was similarly afflicted? Even if it were not, it would need to be replaced before they could come home; the crew of Soyuz 34 had been due to bring them a new ferry and go home in the older one, but this was now in doubt. In the end it was decided to launch Soyuz 34 unmanned, and use the docked Soyuz 32 to return some samples and experiment results to Earth, also unmanned. The crew then swapped Soyuz 34 to the front port to again allow Progress dockings. The dramas for the resident crew were not yet over. The 10 m diameter KRT-10 radio telescope antenna, which had been deployed from the rear port, became entangled with a fixture on the hull when the crew attempted to jettison it. Therefore, on 15 August, the crew ventured outside to cut the antenna free, doing so with little difficulty. While outside, they and also retrieved sample cassettes from the hull of the station. On 19 August 1979, the crew climbed aboard Soyuz 34 and came home having spent 175 days on board Salyut 6.

Soyuz 32 crew

Remarkably, the crew for the fourth expedition consisted of Leonid Popov and, making two flights in a row, Valery Ryumin. Ryumin was a last minute replacement for Valentin Lebedev, who had injured a knee shortly before launch. So it was that Ryumin found himself back on board Salyut 6 on 10 April 1980 reading the note that he had left for the next long-duration crew! In contrast to the previous mission, this crew entertained four visiting crews, three Inter-Kosmos and one carrying out the first manned test of the new Soyuz-T spacecraft. On 11 October the main expedition landed safely back on Earth after a mission lasting a record breaking 185 days. This meant that Ryumin had spent 360 days in space, making him the most traveled cosmonaut or astronaut at that time. He would fly again, but not until 1998, and on board a U.S. space shuttle to visit a Russian space station, a joint mission that would never have been predicted in the cold war days of 1980.

The final expedition to Salyut 6 began on 12 March 1981 when Soyuz-T 4 was launched with the crew of Vladimir Kovalyonok and Viktor Savinikh. This was after Soyuz-T 3 had flown a short three-man mission to the vacant station, both to replace some of the systems of Salyut 6, and to verify the three-man capability of the new Soyuz-T. The main expedition was to last for 74 days and receive two visiting Interkosmos crews, both using the older Soyuz spacecraft. The main expedition undocked and landed on 26 May 1981, closing the chapter on the fantastically

Soyuz-T 4 crew

successful Salyut 6 station. Later that same year, Cosmos 1267, which had been in orbit since April, docked with the forward port. This helped to prove to engineers the concept of expanding future stations with separately launched modules. Cosmos 1267 was, in fact, a remnant of the Almaz program, as it was one of Vladimir Chelomei's TKS designs that had been launched on an autonomous mission lasting 57 days before it docked with Salyut 6.

Salyut 6 had been occupied by five long-duration crews for a total of 684 days; it had also been visited 11 times by short-duration crews, 9 of which carried international crewmembers. Salyut 6 was finally de-orbited on 29 July 1982 after four years and ten months in Earth orbit.

Salyut 6 in orbit

Inside Salyut 6

7

Salyut 7 and Spacelab

Salyut 6 had been an impressive step forward in space station technology and operations. Salyut 7 was the back-up to Salyut 6 and, therefore, similar in design, but it did have improved systems, and extra comforts for the crews that were likely to be aboard for much longer periods than previously. Personal selection of food items was allowed for the first time, and there was a small refrigerator for the fresh food delivered by Progress. Extra storage was provided, but it would prove to be still not enough as the life of the station lengthened.

Launched on 19 April 1982, Salyut 7 was to have a long life and more resident crews than Salyut 6. The introduction of the updated Soyuz-T spacecraft would allow more flexibility in crew visits owing to its ability to spend more time in space. It was also hoped to achieve the first operational rotation of crews, with a new crew arriving and having the station handed over to them before the old resident crew left. This would save considerable time and resources, as it meant that the station would not have to be powered down and up again by subsequent crews.

Soyuz-T 5 was launched on 13 May 1982 with the first resident crew of Valentin Lebedev and Anatoli Berezovoi—this was defined as the EO-1 crew. They were initially given light duties for their first few days in orbit as they worked their way through the tasks required to commission the new station. New experiments were set up, and the crew slowly settled into a daily routine as they awaited their first visitors. Soyuz-T 6 was launched just over a month later, and carried the first crewmember from outside the Inter-Kosmos organization, Frenchman Jean-Loup Chrétien, who was to carry out a series of medical experiments. As did short-term visitors, he wore himself out, shortening his sleep periods to maximize his time in orbit, and by the time Soyuz-T 6 undocked from Salyut 7 to return to Earth, he was exhausted. However, the resident crew of Lebedev and Berezovoi were just as tired, because hosting visitors was hard work, as previous crews had found, and ground controllers gave them a few days off to allow them time to recover. In addition, the two men did not really get on that well; they had not bonded during training for the mission, but for some reason

Soyuz-T 5 crew

they had not been reassigned to separate missions. Two men aboard a small space station is never going to be an easy period of time to get through, particularly when it is for such a long period of time, but this pair seemed to exploit every excuse for arguing with one another, even over trivial things. The only break for the crew came

Soyuz-T 7 crew

when visitors arrived, and the next set of visitors would be more welcome than most, because it included a woman.

Soyuz-T 7 established another space triumph of sorts for the Soviets. Svetlana Savitskaya was the second woman in space after Valentina Tereshkova in 1963. It was obviously no coincidence that NASA had announced earlier in 1982 that Sally Ride would fly on board the space shuttle's seventh mission. The Soviets wished to trump NASA's latest public relations scoop, and assigned Savitskaya to the flight at relatively short notice. However, it is unlikely that the more liberal American Sally Ride would have accepted the flowers and floral apron that were presented to Savitskaya by her male colleagues. Her flight was relatively short, lasting only seven days before the visitors returned to Earth in the older Soyuz-T 5, leaving Soyuz-T 7 for the long-duration crew.

Alone again, the two men struggled to get along; there was a momentary panic when Berezovoi felt unwell one day during an exercise period. His illness threatened the length of the mission, and both men felt angry that having put up with each other for all this time, they might have to come home early. Ground controllers recommended that Berezovoi be given an injection of atropine to ease the pain, and this helped, causing him to feel much better by the next day; the mission could continue. Finally the crew had reached their personal limits, and they were allowed to return home. They had set a new endurance record of 211 days, but their landing and recovery did not go completely smoothly, as they had to spend the night on board a disabled, and cold, helicopter. This was the last straw for the two men, and in the twenty odd years since their joint flight, they have barely spoken to each other.

The launch of Soyuz-T 8 on the 20 April 1983 did not go entirely to plan. The crew of Aleksandr Serebrov, Gennady Strekalov, and Vladimir Titov were unable to dock with Salyut 7 because one of the spacecraft's rendezvous antennas was damaged at launch; they returned to Earth on the 22 April. Soyuz-T 9 docked with the station on the 28 June carrying Vladimir Lyakhov and Aleksandr Aleksandrov. As the next long-duration crew, EO-2, they were due to receive visitors, but unfortunately the launch of Soyuz-T 10-A, again crewed by Strekalov and Titov, was aborted and the launch escape system used when the booster caught fire during the last moments of the countdown. Thus, Strekalov and Titov failed for the second time that year to get to Salyut 7, where they were supposed to add solar arrays to the station. This task would now fall to the resident crew. Following on from the success of Cosmos 1267 with Salyut 6, Cosmos 1443 had docked with the station prior to the arrival of the Soyuz-T 9 crew, and was loaded with 3.5 tonnes of supplies. During its stay, Cosmos 1443 was used to provide attitude control for the station, and to boost Salyut 7's orbit. The re-entry module would later turn up at a Southerby's auction in 1993. The crew set about unloading just after they arrived; they then loaded the TKS' re-entry module with experiment results, which returned to Earth in August. They carried out the spacewalks to install the solar panels (which were cargo in the large module) just a few weeks before their return to Earth on 23 November, after 150 days in space, having received no visitors. At around this time it was noticed by the resident crew, and ground controllers, that Salyut 7 was leaking fuel from its propellant tanks, severely limiting the station's maneuvrability. Plans were made for the next crew to attempt to fix the problem, rather than abandon Salyut 7 at this early stage.

SPACELAB

Having deciding to concentrate on the space shuttle program after the three visits to Skylab NASA lacked a space station of its own. However, in collaboration with the European Space Agency (ESA), NASA developed the Spacelab. This made available a pick and mix of a pressurized module and open pallets that sat in the shuttle payload bay to allow scientific experiments for the duration of a shuttle mission. It was obviously nowhere near as good as the long-duration experiments that could be carried out aboard the Salyut stations, but it was the closest thing possible with the space shuttle. Critics pointed out that it was impossible to make the shuttle a completely gravity-free environment, as the movements of the relatively large crew, plus thrusters firings, would interfere with the results of many experiments. The project began in 1973 when NASA and ESA signed an agreement that outlined the components and responsibilities of the Spacelab project. The first engineering model of a pallet arrived at NASA in 1980, and went on to be used on the shuttle's second flight in 1981. Most Spacelab missions could only last up to 10 days, but NASA added the Extended Duration Orbiter (EDO) pallet to the shuttle and in 1992 STS-50, a Spacelab mission on Columbia, flew a 13-day mission. The longest shuttle mission, STS-80

STS-9 crew

lasted for almost 18 days, and this represented the limit of the shuttle's duration. In total twenty-four Spacelab missions would be flown on the shuttle, seventeen of them with the pressurized lab module, the first of which, STS-9, was launched on 28 November 1983 and lasted for 10 days.

Whilst not strictly speaking a space station component, Spacelab did shape the way NASA planned and undertook its science based missions. The crew's schedule for these missions was extremely tight, with not a minute wasted; of course on a short mission with around the clock shifts of crew members it is acceptable and sensible to plan this way, but it would do nothing to help NASA plan for future space station operations, when it simply would not be possible to plan every last minute of the day.

The Soviets made maximum use of their new ferry craft capabilities on the 8 February 1984 with the launch of Soyuz-T 10. This time the crew numbered three due to the inclusion of a physician, Dr. Oleg Atkov, who would monitor the long-duration crew (EO-3) of Leonid Kizim and Vladimir Solovyov during their record attempt. Kizim and Solovyov had been trained for several EVAs to attempt to fix the leaking fuel tanks. Eventually they would carry out a record six spacewalks in their efforts to fix the leaks and add solar arrays to the station. Salyut 7's future had been assured by the skillful efforts of the cosmonauts and wisdom of the planners on the ground. Two crews of visiting cosmonauts included the first Indian in space Rakesh Sharma, and the return of Svetlana Savitskaya who would make a spacewalk this

Soyuz-T 13, Salyut 7 repair crew

time. Savitskaya was accompanied by Buran chief test pilot Igor Volk, who was using this flight to test a home coming Buran pilot's ability to land his craft on a runway at the end of a long flight. Upon landing on 29 July, Volk immediately flew a MiG fighter to 21 km before landing with dead engines to simulate a Buran landing. The three-man EO-3 crew landed on the 2 October having set a new duration record of 237 days in space, which would be the longest single-crew stay aboard Salyut 7. Vladimir Dzhanibekov, who had commanded the Soyuz-T 12 mission with Savitskaya and Volk, could not have had any idea that he would be returning to the station in less than a year, or why.

The year 1985 was to be a somewhat more complicated and dramatic year for Salyut 7 and its crews. It began when Mission Control lost all contact with the station on 11 February; it had lost all attitude control and had gone into free drift mode, making it impossible for a Soyuz ferry to automatically dock with the station. The crew of Soyuz-T 13 were dispatched on 6 June with Vladimir Dzhanibekov and Viktor Savinykh to try and determine what had gone wrong. When they rendezvoused, the station appeared to be undamaged, although it was clearly without power, there being no lights, and the solar arrays pointing in differing directions. The station was slowly rolling around its long axis, but Dzhanibekov was able to line up the Soyuz with the aid of docking controls that had been installed in the orbital module for just such a purpose. They managed to dock, and entered the dead station; it was dark and cold as it had been completely powered off. By the crews own crude estimate, the interior temperature was about $-10°C$, an estimate reached by spitting on the bulkhead and timing how long it took to freeze! Clearly, they would have to wrap up to work in these conditions, and return intermittently to the Soyuz to warm up. To attempt to bring the station back to life, the crew fitted spare batteries, replacing the existing ones that would not charge back up. In the process of this work they discovered a faulty charge sensor. This sensor determined if a battery was full or in need of charging, and it had failed in such a way that the computer thought that all of the batteries were fully charged and stopped trying to charge them; as a result all of the batteries went flat, and the station died. If a crew had been on board, the faulty sensor would have been immediately detected, and replaced well before the station lost all power. Once this sensor was replaced, the task of recharging the batteries began, and the station slowly came back to life. The crew had saved the station, once again proving the value of humans in space, and proving that the Soviets were now very comfortable with repairing their spacecraft, rather than just launching new ones when something went wrong. A fact that they would be keen to underline when failures began to undermine the fledgling partnership with NASA.

Soyuz-T 14 arrived on 18 September with Georgi Grechko, Vladimir Vasyutin, and Aleksandr Volkov aboard. Vasyutin and Volkov had trained with Savinykh as the original long-duration (EO-4) crew, so when Soyuz-T 13 landed on 26 September it left behind the EO-4 crew to begin their mission. Unfortunately, during October Vasyutin became very ill; his temperature was very high (about $40°C$), and the ground advised him to rest in the hope that the fever would pass. It did not get any better; in fact he seemed to get worse, and Valeriy Ryumin ordered an immediate end to the mission. In actual fact, it took the crew about a week to prepare the station for autonomous flight and return to Earth, by which time Vasyutin had become very ill indeed. Upon his return he was immediately taken to hospital, where he took a month to recover from what turned out to be a prostate infection. It was an unfortunate end to a promising long-duration mission by Savinykh, who was very disappointed to have missed the duration record. It was also unfortunate for the future of Salyut 7, which had clearly reached the end of its useful life. The rescue mission had also used a Soyuz that was to have been utilized by an all female crew commanded by Svetlana Savitskaya with two flight engineers Yekaterina Ivanova and Yelena Dobrokvashina. After the cancelation of their flight it was hoped that they might fly to Mir, but Savitskaya became pregnant in 1986, and the idea was abandoned. Ivanova and Dobrokvashina were never assigned to another mission, and both left the cosmonaut corps in 1993.

The EO-4 mission was to be the last planned long-duration flight to Salyut 7; its successor Mir had been launched on the 19 February 1986, and it seemed as if Salyut 7's operational life was over. However, a unique mission was planned that would see the crew of Soyuz-T 15, Leonid Kizim and Vladimir Solovyov, activating the new Mir station, and then flying their Soyuz to dock with Salyut 7 to complete and collect the work not finished by the EO-4 crew. So on 5 May they undocked from Mir after six weeks aboard and transferred to Salyut 7 the next day. After 50 days aboard Salyut 7 they returned to Mir for a further 25 days before returning to Earth on the 16 July after a truly unique mission.

Salyut 7 in orbit

Salyut 7 stayed in orbit until 7 February 1991 when it re-entered the atmosphere and was destroyed. The stage, however, had been set, for Mir was now operational and offered much more flexibility than the previous Salyut stations. The best was yet to come.

Color plates

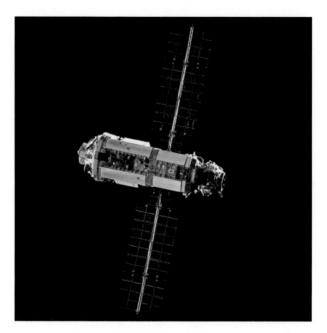

Zarya after launch

Zarya and Unity after STS-88

Zarya, Unity, and Zvezda

ISS after STS-97 adds the first solar arrays

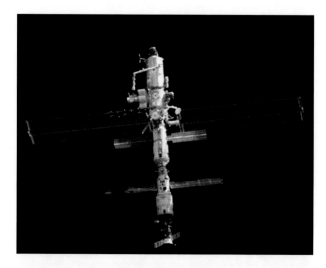

ISS after the Destiny lab and Quest airlock were installed

ISS after the STS-114 return to flight mission

ISS after STS-116 added more solar arrays

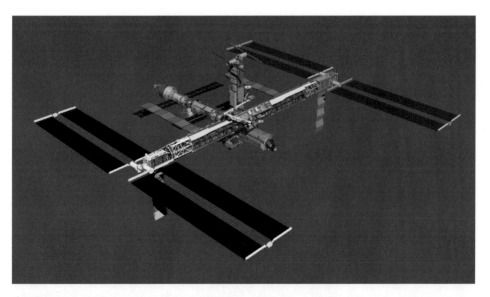

ISS after STS-117

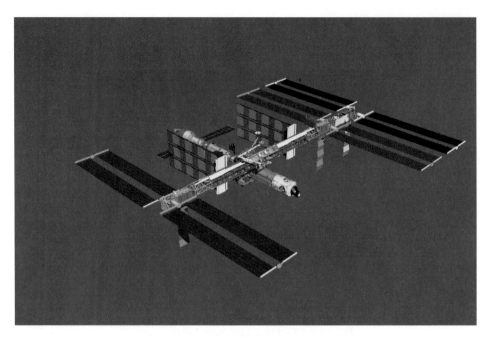

ISS after STS-118 and STS-120

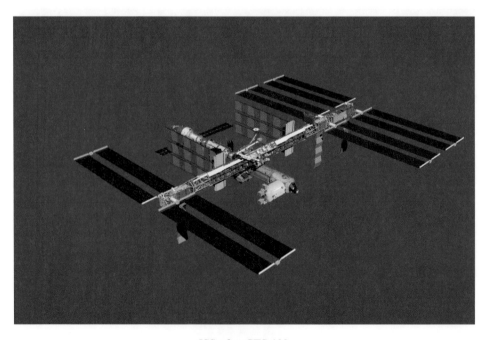

ISS after STS-122

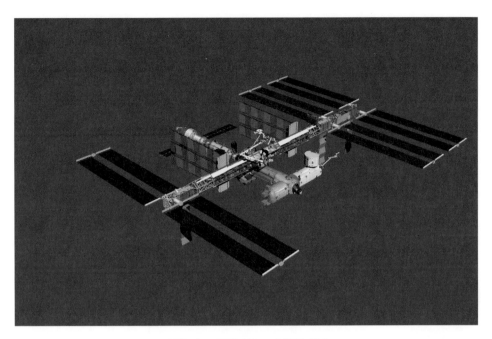

ISS after STS-123 and STS-124

ISS after STS-119

ISS after Node 3 is attached

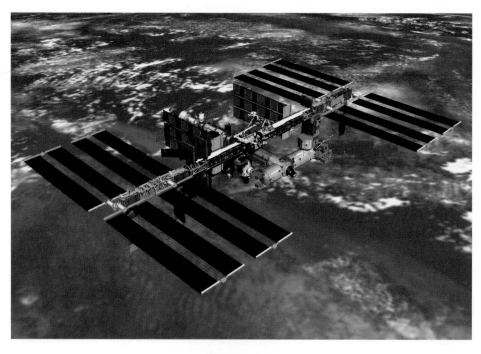

ISS complete

8

Mir: For all mankind?

The very name Mir seems to conjure images of disaster, and words like beleaguered and trouble-torn were usually associated with it, for this was the only way that this outstanding space station was ever mentioned in the popular news programs and newspapers. This image was reinforced in popular culture by Mir's depiction in movies such as "Armageddon". The truth, of course, was somewhat different; the facts are simple, Mir was in orbit for 15 years, and played host to over 100 cosmonauts and astronauts. It is true that in later years it required more maintenance than in its earlier years, most things do, but its legacy will stand for many years to come. The incidents that led to Mir's unfortunate reputation are described in Chapter 10.

The name Mir is variously translated, but can mean "peace", "community", or "new world"; but perhaps most significant was the fact that it had a name at all, as opposed to being referred to as "Salyut 8". However, it soon became clear that this station was meant to be a new beginning for Soviet space stations, with a long life planned for it. Mir would embody everything that had been learned previously, and hence with a new beginning came a new name. It did not hurt that the new name would strike a welcome cord with the new General Secretary of the Soviet Union, Mikhail Gorbachev.

Unusually, Mir was launched whilst its predecessor Salyut 7 was still in orbit, raising speculation that some kind of joint operations were intended, and maybe even a docking between the two. Its launch in February 1986, barely a month after the hammer blow of the Challenger launch disaster, highlighted the Soviet Union's relentless presence in space, and seemed to press home, cruelly, its continued progress in long-term space flight.

Mir was different from the earlier Salyut stations in an important way. Its most important addition was the four docking ports arranged around the radial axis of the front end. These would allow the station to be expanded with science modules. This, in turn, meant that the core module or base block, as it was known, had more space; it was primarily a habitat module for the two or three permanent crew. The stations

Soyuz-T 15 crew

solar panels were larger than those on Salyut 7, and more panels were to be fitted shortly by spacewalking cosmonauts. The computers on board Mir were sufficiently advanced as to allow the crew more time for scientific activities; in fact, the whole station's design reflected the fact that this station was meant to last longer than any of its predecessors.

Mir was to be activated by the crew of Soyuz-T 15, who were launched just a month after Mir was established in orbit. Two experienced cosmonauts, commander Leonid Kizim and engineer Vladimir Solovyov, were selected to not only carry out the first mission on Mir, but also to visit Salyut 7 and finish the outstanding experiments on that station. Once they had rendezvoused and docked with Mir, the crew found a much roomier cabin than the previous Salyut stations, which both crew-members had spent considerable time aboard. Although the physical dimensions of Mir's base block were about the same as previous stations, the interior was much less cluttered—a reflection of the plan to add modules later for scientific research. For the first time the crew had their own individual cabins, with sleeping bag, window, and storage for personal items. The bathroom offered some privacy, and a kind of wash basin, and the table at which the crew would eat was a great improvement over earlier facilities. In all, Mir was designed with long-duration space flight in mind, and offered a level of comfort not seen on a space station since Skylab. The lessons learnt from previous station operations was also evident in the plan for the working day; it would follow a more usual five days a week schedule—with a normal working day's duration and with time of in the evening for the crew to relax or pursue their own interests. The crews were also left free to determine their own schedules for the day; a marked difference from NASA's "plan every minute" approach to space flight. The Russians seemed to understand that long-duration missions were like running a marathon; the crew had to pace themselves to keep their efficiency levels up as well as their spirits.

Kizim and Solovyov spent the next several days preparing Mir for its mission; they unpacked an already docked Progress, and generally readied Mir for long-term space flight. One Progress left and another arrived to continue the process of activation, and to ensure that Mir's propellant supplies were topped up. As the beginning of May approached, the crew put Mir back into an autonomous operating mode; they were leaving, but not for good, they were going to Salyut 7. Transfer between two orbiting space stations had never been achieved before, or since. On 5 May 1986 Soyuz-T 15 undocked from Mir to begin the one-day transfer to Salyut 7, docking with the veteran station was easily achieved, and in fact the whole process was made to look routine. The plan was to activate Salyut 7 once more, and finish off the remaining experiments on board the station. Toward the end of the month, the crew ventured outside Salyut 7 for the first of two spacewalks to retrieve a number of external experiments and to test the deployment mechanism for a structure that would eventually be built on Mir. By the end of June the crew was ready to return Salyut 7 to solo flight, and take as much equipment back to Mir as they could pack into the orbital module of their Soyuz; they had been on board Salyut 7 for 50 days. After a trouble-free return trip to Mir, the crew settled into a routine once more, concentrating on installing the equipment transferred from Salyut 7, and on their exercise regimes in preparation for the return to Earth. It had been assumed that the crew would hand over in orbit to the next, but apparently the next crew were not yet ready, and in truth Kizim and Solovyov had run out of things to do. On 16 July they landed after an historic and successful mission that had seen them occupy two space stations for a total of 125 days.

In fact it was some time before Mir was to be occupied again. The first expansion module for Mir, called Kvant, had suffered a few delays as it was modified from its original design as an adjunct to Salyut 7. There had also been delays with the crew, originally scheduled to consist of Vladimir Titov and Aleksandr Serebrov, when Serebrov failed a medical exam they had to be replaced by their back-ups Yuri Romanenko and Aleksandr Laveikin. Titov did not seem to be a lucky man; so far his career had consisted of a failed docking attempt with Salyut 7, and the launch pad abort, and, now he had been removed from a mission through no fault of his own. Many of his cosmonaut colleagues wondered if he was cursed.

When the crew did launch on 6 February 1987, it did so on board an upgraded Soyuz design with features specifically designed for the new orbital outpost. The Soyuz-TM was a necessary upgrade to the existing Soyuz-T craft because of the new rendezvous system used by Mir called Kurs. This new system basically allowed the Soyuz to dock automatically without Mir having to change its own orientation; a great saving of the limited maneuvring fuel available on the station. In addition a new window had been added to the orbital module to allow a crewmember to directly view the upcoming docking, and the interior of both modules had been slimmed down to save weight and give the crew more space.

Yuri Romanenko and Aleksandr Laveikin arrived at the station on 7 February 1987, docking with Mir's front port because a Progress cargo craft was already at the

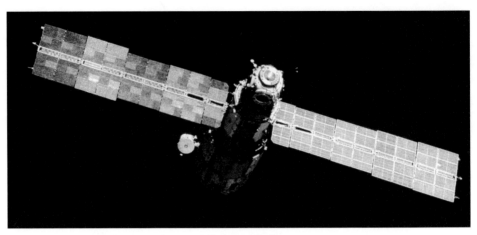

Mir base block

rear port. It took some time for Laveikin to adapt to life in space; it was his first flight, and he said that it took the best part of a month to feel comfortable in orbit. Romanenko had no such difficulties, he had flown before, spending three months on Salyut 6, and adapted readily to the new station. Once the new crew had settled in they waited for the new module to be launched.

The first laboratory module, Kvant, was launched on 31 March 1987. As it had no propulsion system of its own, it was mated to a modified TKS serving as a tug. The tug was to deliver Kvant to its automatic docking with the rear port of Mir, its permanent home. Kvant made its first docking attempt on 5 April, but something went wrong and the module sailed past the station, with a somewhat concerned crew watching it pass Mir's portholes. A second attempt a few days later achieved only a soft docking; when the docking probe was retracted the latches failed to lock. It was decided to get the crew to go outside and have a look. So on 11 April they ventured out and found a cloth bag full of hygiene towels that had somehow escaped from the previous Progress craft—it had blocked the hard docking, which was achieved successfully once this object was removed. The crew entered Kvant for the first time on 12 April for an initial inspection. The interior consisted mainly of equipment for an electrophoresis system for processing biological materials, and there was also substantial equipment for carrying out astrophysics observations. In addition to the experimentation equipment, there were additional devices to help with the operation of the station in general. Elektron took water (whether reclaimed vapor, waste water, or urine) and electrolyzed it into oxygen and hydrogen—the oxygen for the life support system and the hydrogen vented into space. Another very important piece of operational equipment were the stations gyrodynes; these spinning flywheels were used to rotate the station as required, rather than using valuable propellant via the thrusters. The future expansion of Mir had originally been planned around the use of more Kvant sized modules, but at some point it had been decided to concentrate on modules more than twice the size at around 20 tonnes each, based on the TKS design.

ENERGIA FLYS WITH POLYUS

On the 15 May 1997 the Soviet Union achieved something that had eluded it for many years, the launch of a heavy lift booster. As we have seen in earlier chapters the ill-fated N-1 moon rocket endured four failures before its cancelation, but the brand new Energia rocket was launched successfully first time. The payload for its maiden launch seemed a very simple one at first glance. The Soviets reported that it was a mock-up of a manufacturing and material processing platform known as Polyus, future versions of which would be used either as add-on modules for existing space stations, or as free-flying platforms for particular missions. It had a mass of about 80 tonnes and was slightly larger than the existing Mir base block. Unfortunately, in

Polyus on first Energia—note "Mir-2" written on the side

this case the platform did not perform as designed; whilst the Energia rocket performed perfectly, the payload fired its own insertion engine at the wrong orientation and propelled itself back toward the Earth, destroying itself in the process.

Equally unfortunately, all of the above description of the payload from the Soviets was totally inaccurate. Polyus was indeed its name, and it did weigh about 80 tonnes, but in actual fact Polyus was a military orbital weapons platform prototype, a system that apparently Soviet Premier Gorbachev had ordered not to be launched in order not to jeopardize his delicate negotiations with U.S. President Reagan. Basically, Polyus was the Soviet's response to Reagan's "Star Wars" Strategic Defense Initiative. It consisted of many pre-existing space components like a TKS tug, which was similar in design to the FGB or Functional Cargo Block that would be launched as Zarya, the first component of the International Space Station (ISS) many years later. It was also thought to include defensive armaments, and test targets that could be released to test its on-board weaponry. None of these features were ever confirmed, and in fact very little information on this "battle platform" has ever come to light. The answers lie at the bottom of the Pacific Ocean for anyone that wishes to look.

MORE ADDITIONS TO MIR

The delivery of Kvant allowed some cargo to be brought to Mir as well; one of the items stuffed inside the new laboratory was a new set of solar arrays that the crew would locate on the base block of Mir. It arrived in two sections, so the crew would need to venture outside twice to finish the work. During the previous Extravehicular Activity (EVA), physicians on the ground had noticed irregularities in Laveikin's heart rhythm which caused them some concern; however, after further studies it was decided to allow him to carry out installation assembly. The two spacewalks were carried out without incident, and the installation added about 2.5 kW of power to the station's total supply. Unfortunately, the spacewalks allowed the doctors on the ground to further study Laveikin's heart, and they came to the conclusion that they could not tell enough from the remote telemetry to perform a proper diagnosis. Poor Laveikin was told to try and relax, and a plan was put in place to get him home as soon as possible. The next crew would arrive in July, and would consist of Aleksandr Viktorenko, Aleksandr Aleksandrov, and a Syrian visitor, Mohammed Faris. In order to relieve Laveikin, Alexandrov would replace him on the permanent crew and Laveikin would return to Earth on board Soyuz-TM 2 with Viktorenko and Faris in July 1987.

Romanenko had now been in space for the best part of a year, and the end of his flight was approaching. He had hoped to increase the endurance record to a full year, but his increasing testiness with the ground and his crewmates convinced Soviet officials to bring him home short of his goal. He returned to Earth on board Soyuz-TM 3 with Aleksandr Aleksandrov and Anatoli Levchenko who had been launched on a taxi mission toward the end of December. Levchenko was another member of the Buran test pilot group and he, like Igor Volk, was flying to check the landing ability of a cosmonaut after exposure to weightlessness. Immediately after landing he was flown by helicopter to a Tu-154 civil airliner which he used to simulate Buran landings. The Buran program was in crisis, however, and faced cancelation at any moment. Despite this fact, Buran made its first flight, unmanned, in November 1988, completing one orbit before returning to the launch site under remote control, accompanied by a MiG-25 chase plane flown by Igor Volk. This flight was a significant achievement (the U.S. shuttle could not be flown unmanned, as it at least required a crew member to lower the landing gear) but it had come too late to save the program. The planned flight to Mir, which was scheduled for December 1994, was canceled, and the program itself was concluded in June 1993 by Boris Yeltsin.

It would fall to the very next crew to break the one year in space barrier. Vladimir Titov, finally breaking his jinx, and Musa Manarov, a rookie cosmonaut, were to fly to Mir on board Soyuz-TM 4 with Levchenko, and return to Earth 365 days later on Soyuz-TM 6. At the end of their marathon flight, Dr. Valeri Polyakov would begin the first of his record-breaking flights. The timing of his flight was important; he wanted the chance to observe Titov and Manarov whilst they were still in space to see for himself the medical effects of such a long flight, before he began one of his own. His mission continued well, and when Titov and Manarov returned to Earth he was joined by Aleksandr Volkov and Sergei Krikalev who, it was planned, he would finish the long-duration mission with. However, on the ground things were not proceeding so well. Volkov and Krikalev had been trained to receive the next new modules to expand the Mir complex, but the construction of those modules had slowed to a crawl; there simply was not the money to complete them, let alone launch them. Finally it was decided to bring the crew home and mothball the station for the next five months. Polyakov was devastated, despite having spent 240 days in orbit, and upon his landing immediately began canvassing for another, longer mission.

Aleksandr Viktorenko and Aleksandr Serebrov were the next occupants of Mir; they launched on board Soyuz TM-8 on 5 September 1989, and entered the station a few days later. The second Mir expansion module was now ready after the delays, and on 26 November it was launched to dock at Mir's front port. The flight to Mir was not entirely smooth; one solar array initially failed to deploy, but was shaken loose by putting the module into a slow roll. Even when it arrived at the station, its Kurs automatic docking system aborted the first docking attempt; however a second docking four days later was successful. Kvant 2, as it was known, then swung itself to an upper docking port by using its Ljappa "swing arm" to rotate itself 90 degrees before reattaching itself to Mir. When the cosmonauts entered the new module they found three new compartments. The nearest compartment gave the cosmonauts some new home comforts, a shower, and a second toilet. The middle area contained room for scientific experiments, but could also be used as a back-up airlock. The compartment at the end of the module was intended to be Mir's main airlock; it had enough room to store extra spacesuits, and a wide outward opening. One of the reasons for the wider hatch was also contained in the area, the Soviet version of a manned maneuvring unit, called Icarus. The first test of the Icarus unit, which was intended to be used later with the Buran space shuttle, would be slightly different from the one undertaken by NASA's Bruce McCandless and Bob Stewart on the shuttle mission

STS-41B five years earlier. Serebrov and Viktorenko remained tethered to the station throughout their test, and whilst McCandless and Stewart had been able to fly free of the shuttle up to distances of 320 ft, the cosmonauts never reached more than 150 ft. The reason for this, of course, was the fact that if the unit had failed, the shuttle would have been able to go and collect McCandless or Stewart; Mir could not really go anywhere to retrieve anyone. This would prove to be the first and last use of the Icarus system, it was thought to be too complicated and risky to use, and eventually was left outside the station to free up space in the airlock.

The next new Mir module, Kristall, arrived at the station in June of 1990. Once again the first docking was aborted by the automatic system, and again the second attempt was successful. In the same manner as Kvant 2 before it, Kristall's own small robot arm moved the module to its dedicated docking port on the node, directly opposite Kvant 2, giving the station a "T" shape. Kristall's interior was very different from Kvant 2; it was mostly fitted out with furnaces to allow metallurgy and crystal-growth experiments. At the far end of the module was a docking unit to allow the Soviet shuttle Buran to dock; this would never be used by Buran, but it would be used once by a suitably fitted out Soyuz, and then much later and ironically, by the U.S. space shuttle. Although both of the new modules added significant internal volume to

Mir as it appeared in 1993

Soyuz-TM 12 crew, with British astronaut Helen Sharman

the station, it was still a great deal less than Skylab had provided its crews, and the total weight of the complex was about 12 tons less than Skylab too.

Mir now settled into a period of long-duration missions intermingled with visits by international cosmonauts. Visitors from Japan, Great Britain, Austria, France, and the newly reunified Germany took place over the next four years.

The most significant long-duration missions were undertaken by Sergei Krikalev who spent 311 days in 1991/2 in addition to the 151 days that he had accumulated in 1988/9. Whilst he was there, the world below him changed. He was launched as a Soviet citizen, but the revolution that caused the Soviet Union to collapse also returned him to Earth as a Russian citizen. He was dubbed "The Last Soviet Citizen" by the press. He would go on to fly two more mission to the ISS, to bring his total time spent in space to 803 days; over 50 more than the previous record holder. At the time of writing Krikalev is due to return to the ISS as commander of Expedition 19 in March 2009.

However, the man to have spent the most time in orbit in a single mission is Dr. Valeri Polyakov. In January 1994 he returned to the station with Viktor Afanasyev and Yury Usachev on Soyuz-TM 18. Polyakov had managed to sell the idea of an ultra-long-duration flight to the space program officials on the basis that it would attract new international interest in joint missions, interest that would

Dr. Valeri Polyakov

bring much needed currency to the now Russian space program. On his arrival, he was welcomed by Aleksandr Serebrov and Vasily Tsibliyev, who were due to return to Earth a few days later. Tsibliyev was on his first space flight, and Serebrov his last after four missions. When they undocked from the station, Tsibliyev flew a fly-by to take photographs of Mir. Unfortunately, the Soyuz gave the Kristall module a glancing blow. Worse collisions were to come in later years for both Tsibliyev and Mir, but for now no damage was done, and the spacecraft returned to Earth without further incident. Polyakov could be forgiven for thinking that his much-wished-for mission might be over before it hardly began, but no damage had been done to the station either, and the mission proceeded. His stay, beyond being almost indescribably long, was also uneventful, and on 22 March 1995 he climbed into Soyuz-TM 20 along with Aleksandr Viktorenko and Yelena Kondakova to return home. He had spent almost 438 days in orbit, and when his capsule landed in Kazakhstan he walked from it to a nearby chair, a tremendous achievement. He also stole a cigarette from a friend nearby, but could hardly be blamed for that. He sipped a small brandy and inwardly celebrated his mission. His record still stands today, and it is unlikely to be broken until man ventures to Mars.

Altogether, 28 "main expeditions" worked aboard Mir, and they were visited by many short-term crews. A total of 104 men and women visited Mir, including 42 Soviet or Russian citizens. The remainder comprised 44 from America, 5 from France, 3 from the European Space Agency, 2 from Germany, and one each from Syria, Bulgaria, Afghanistan, Japan, the United Kingdom, Austria, Slovakia, and Canada.

On 23 March 2001, Mir was de-orbited over the Pacific Ocean, with any hardware that survived the entry process falling harmlessly into the sea. It was truly the end of a remarkable era; for 15 years Mir had orbited the Earth, and whilst in its final years it may not have been pretty, it was the greatest single achievement yet in the history of manned spaceflight.

9

Freedom: The U.S. strikes back

NASA had desired a space station since the demise of Skylab in 1979, but the financial and technical constraints of the space shuttle program had made such an undertaking impossible. As we have already seen, the Soviet Union had made great strides in space technology and usability, and were far ahead of the Americans in this area of manned spaceflight. NASA was eager to use the space shuttle to gain back some of the ground that had been lost.

Many attempts had been made by the then NASA Administrator James Beggs to persuade the President and Congress to fund development of a new space station but he had always been unsuccessful. Despite this fact, in 1982 NASA went ahead and obtained eight different designs from the big aerospace contractors of the time, hoping that one of them would finally convince Congress of the value of a new station. Most of the contractors, however, came up with designs geared toward servicing and launching spacecraft rather than purely scientific research stations.

Finally, in January 1984, despite great opposition from some of his advisors, President Ronald Reagan announced the new space station in his State of the Union address, and directed NASA to assemble it within a decade. International partners such as Canada, Japan, and the European Space Agency (ESA) would provide hardware for the station, as well as technical support. NASA was to keep the first two years as a low-key definitions program in order not to incite the many scientists and military leaders who were against the project.

NASA had narrowed down the design options to four by March 1984, and the main baseline configuration chosen was the "Power Tower" design which had been submitted by Boeing/Grumman. The main reason for this choice was that it allowed the most flexibility for future expansion without adversely changing the stations overall mass; it kept NASA's options open. The Power Tower provided a clear area for shuttle dockings, as well as predefined attachments for specific temporary pay-loads. It was thought that the entire assembly could be carried out by 12 shuttle launches over a 3-year period, but other contractors doubted this. In late 1985 NASA

"We can follow our dreams to distant stars, living and working in space for peaceful, economic and scientific gain. Tonight, I am directing NASA to develop a permanently manned space station and to do it within a decade."

President Ronald Reagan
State of the Union Message
January 25, 1984

Reagan gives state of the union message

Space station design in January 1984 (purely illustrative)

Power Tower

Rockwell "Dual Keel" design 1985

changed its baseline configuration, and abandoned the "Power Tower" concept that it had already spent a considerable amount of money on due to complaints from potential crews and engineers who felt that the design would not prove stable enough for scientific experiments. The "Dual Keel" design became the new baseline. It was based on Lockheed and McDonnell-Douglas designs, and was chosen because it was felt that it would provide a much stiffer structure, and therefore a better microgravity environment for experimentation. The crew complement was increased to eight to allow more scientific work to be carried out.

In 1988, space station Freedom, as it was now officially known, was going to cost at least $14.5 billion and would require 10 or 11 shuttle launches to complete. And it was felt by many that NASA was playing down the true cost by not including all shuttle launch costs. In addition, there were doubts that the space shuttle could reliably service such a station. The space shuttle was, of course, central to the plans to construct the space station. Its unique capability to carry large payloads into orbit and have a crew on board capable of joining the pieces together meant that literally nothing else could do the job. The United States was seriously lacking an unmanned heavy lift launch vehicle; the shuttle had been imposed on all commercial and military customers as the only game in town. Consequently, the space station and all other launch customers were left in disarray by the disaster that befell the space shuttle Challenger on 28 January 1986. The fact that the accident had largely been caused by NASA's own mismanagement, as well as a flawed booster design, eroded confidence in NASA in other areas, and that included designing, building, launching, assembling, and maintaining a manned space station.

By far the biggest problem, however, was that NASA was trying to please everybody with the space station design. It was trying to offer a garage for assembly of interplanetary spacecraft, a massive variety of scientific laboratory facilities, including animal research, variable gravity, materials processing, life sciences and the like. The power requirements for all of these capabilities were massive, and would need solar arrays of the greatest quality. One station simply could not carry out all of these contradictory requirements; not without being a massively expensive leviathan, which is what it had become. The program was far too large for its own good, and NASA seemed more concerned with pushing the frontiers of technology instead of designing a station that they could actually launch and maintain within a reasonable budget. NASA needed to decide on the station's primary use. Contradictions were caused because, for instance, animal research or spacecraft assembly would adversely affect the microgravity environment needed for materials processing or other scientific experiments.

The situation was not improved by the Department of Defense, which in 1987 demanded full access to the station to carry out military research. NASA's partners were incensed, and the situation had to be quickly resolved to ensure continued involvement. By 1989 the estimated cost had grown again, to $19 billion; and this was after NASA had deleted some capabilities from the station and reduced its power requirements. In addition, a new program to improve the performance of the space shuttles solid rocket motors was required to launch the ever-increasing weight of the station, which increased the overall cost even further. This trend was to continue until 1990, when Congress demanded a major rethink. The existing design, as well as being overweight and a long way over budget, was also going to require far more maintenance once it was built than NASA had planned for. It was estimated that around 3,000 hours of Extravehicular Activity (EVA) work would need to be carried out per year in contrast to NASA's target of around 500 hours. The redesigned station, now nicknamed "Fred" by critics (to indicate that it was a cut down Freedom), was unveiled in March 1991, would cost around $16.9 billion and would take 23 shuttle launches to complete.

By the time the Freedom project was canceled in 1991, NASA had redesigned the station at least six times and spent over $11 billion without building a single piece of flight capable hardware. Valery Ryumin of Energia was heard to comment, "They've spent 10 years and $11 billion; if only we'd had a bit of that money. $11 billion and they haven't done a thing; everything they've done in that decade was useless, none of it worked. Ten years and all they built was a wooden model."

Although canceled in May 1991, the space station plan was quickly revived only one month later; but with a dramatically cut budget. It was not until 1993 that President Bill Clinton really tackled the problem directly. He demanded three new station designs, options A, B, and C, costing $5 billion, $7 billion, and $9 billion

Space station Fred—March 1991

respectively. Option A, which was based on a 1991 Freedom design, was chosen as the best compromise, and would cost $6 billion, but this would be without a habitation module that would be added later at additional cost. Nevertheless, the station's critics in Congress remained skeptical, and a move to kill the entire project failed by a single vote. At this point, NASA introduced a new partner—Russia. Using Russian modules and technology would make the assembly of the station more efficient. Clinton saw an opportunity to tie the Russians into a program that would keep its engineers busy, and therefore less likely to get involved with other countries more questionable activities. It was only when this agreement was reached (Chapter 11) that things began to move forward; mostly because the station now had an acceptable political face.

In reality, the same problems that had plagued Freedom would continue into the ISS. It was never very clear what Freedom or the ISS was actually for. What goals did it set? The Soviets had always had the goal during the many iterations of Salyut to make each station more independent, more self-sustaining, than its predecessor. This kind of technology and operational capability would be necessary for the longer, far-reaching space flights of the future, like a manned mission to Mars. With Mir, Russia had almost achieved the ultimate goal of a "closed loop" spacecraft. However, Freedom and later the ISS would not have the same goal; there was nothing "closed

First ISS design—1993

loop" about the design, and this did not appear to be the goal in the future. When Ronald Reagan made his speech in 1984, he said, "America has always been greatest when we dared to be great. We can reach for greatness again. We can follow our dreams to distant stars, living and working in space for peaceful, economic, and scientific gain." This was not really the clear goal that NASA was looking for or needed, and it was not long before the old engineering maxim, "the better is the enemy of the good", showed itself to be true.

The space station became far too big and complicated; NASA had designed a Rolls Royce when it only really needed a Mini.

10

Shuttle–Mir: Real co-operation

The Shuttle–Mir program was born in July 1991 when President George Bush and Mikhail Gorbachev signed an agreement for a Soviet cosmonaut to fly aboard the space shuttle, and a U.S. astronaut to fly a Soyuz-TM mission. The two great powers had wanted to build on the Apollo–Soyuz mission of 1975 for some time; there had been suggestions that a Soyuz would visit Skylab, but that was deemed unpractical because of the differences in the docking interface between the two craft. There had also even been a suggestion in 1984 that the shuttle dock with Salyut 7 in a sort of simulated space rescue, but nothing ever came of the idea.

This agreement was expanded upon in October 1992 to include a shuttle mission to the Mir space station, and a long-duration stay by a U.S. astronaut on Mir. The mission to Mir by the shuttle was made possible by the availability of the docking adapter that originally had been built for the Soviet shuttle, Buran. That adapter would now be fitted to the shuttle Atlantis. For the first time in its history the shuttle had somewhere to go, although the shuttle's original designers surely had no idea that its first such mission would be to a product of the Soviet Union!

In 1993 it was decided that the shuttle would in fact dock with Mir ten times, exchanging crews and allowing U.S. astronauts several long-duration missions. It was additionally agreed that more than one Russian would fly on the shuttle. By this time, of course, Russia was a partner in the newly redesigned International Space Station (ISS), and so the new program was divided into three phases. Phase 1 would see the ten Mir–Shuttle dockings, involving at least five long-duration flights by NASA astronauts on board the Russian station for which NASA would pay a fee. There would also be at least two flights by Russians on board the shuttle. Phase 2 would signal the beginning of construction of the ISS with launches from the U.S. and Russia of station elements, that would lead to a permanent three-man crew. Phase 3 would complete construction with the elements from other partner nations such as Japan and the ESA.

STS-60 inflight crew portrait

Sergei Krikalev was chosen along with Vladimir Titov as the first Russians to train for a flight on the shuttle as mission specialists. Krikalev became the prime candidate, and he flew on STS-60 in February 1994. This mission had nothing to do with space station operations, but it allowed Krikalev to discover how Americans flew in space, and that flying on the shuttle was very different to flying a Soyuz to the Mir station. This mission lasted a mere eight days; a short sprint in comparison with the months he had spent on board Mir. Activities were far more intense and scripted than his time on Mir, and it proved that the NASA mission planners would have to change their strategy considerably when it came to mounting both the long-term Mir missions and future mission to the ISS.

To underline this message, Vladimir Titov flew on board the shuttle Discovery on mission STS-63 in February 1995. This was not a docking mission, but it was planned to rendezvous with Mir as an engineering demonstration. Although the mission did suffer from some unfortunate malfunctions including one that postponed the flight and another, a thruster leakage, which nearly cancelled the close approach to the Mir station, valuable data for the future docking missions was obtained.

The next part of Phase 1 to be fulfilled was the first long-duration flight by a NASA astronaut to the Mir space station. Additionally this astronaut, Norman Thagard, would be the first to be launched aboard a Soyuz-TM spacecraft. Thagard was an experienced flyer with four shuttle flights under his belt, but he and NASA quickly discovered what Krikalev and Titov had on their shuttle missions, which was

Thagard in his sleep restraint on Mir

that a space shuttle mission and a long-duration flight on a space station are two very different things. The Mir mission required much more flexibility, both from the crewmember and from those on the ground, but NASA seemed to have forgotten the lessons learnt twenty years previously with Skylab. Thagard was dismayed to discover that his ground controllers were programming every minute of his day from waking in the morning to going to bed at night. Much the same frustrations that had plagued the Skylab 4 crew now manifested themselves in Thagard. The problem was exacerbated by the cultural differences for Thagard; he was completely cut off from his compatriots, and often went for days without speaking any English or speaking to his friends and colleagues.

The historic docking of the space shuttle Atlantis with the Mir space station echoed that of the Apollo–Soyuz mission in 1975. Atlantis launched on 27 June 1995 and docked with Mir two days later. The crew consisted of five U.S. astronauts, and the new Mir 19 crew of Anatoli Solovyov and Nikolai Budarin, so that Mir now had a combined crew of 10, the largest in history, beating the previous record of 8 on shuttle mission STS-61A, which flew the first German Spacelab mission. Atlantis also had a Spacelab module in its payload bay to take advantage of the opportunity to study the physiology of the three existing Mir crewmates whilst still in space. Atlantis had brought plenty of supplies to Mir, far more than the Progress freighters could carry, and better, it allowed many things to be returned to Earth from Mir, something that could only be done in small quantities in the Soyuz spacecraft. It allowed the Russians to return faulty equipment to allow diagnostics by Russian engineers; it also allowed experiment results to be returned quickly. Perhaps the most important item delivered by the shuttle was water. Mir was able to recycle about 60% of its own water, but most of that was not fit for drinking. Atlantis was able to deliver half a tonne of water from its fuel cells, where water is a natural by-product and would normally be dumped overboard. It was during all of this back and forth from Mir to Shuttle that it was realized that the ISS would need very careful stock control in order to determine what was on board the station, where it was, and its current status. Mir had not benefited from this kind of control and consequently many items that were unknown, or at least forgotten, were crammed into every available space, often behind wall panels.

Bonnie Dunbar had originally been due to be left behind by Atlantis for a long-duration mission of her own, but this plan had to be abandoned to allow the European Space Agency (ESA) to carry out a long-duration mission by Thomas Reiter of Germany. At this point in time, with the ISS construction running late, many nations wished to fly experiments and people on Mir to gain experience. This meant that finding space on Mir's increasingly busy schedule was difficult. Reiter joined the Mir crew on Soyuz-TM 22 in September 1995, and expected to stay on board for 135 days, although his mission was eventually lengthened by a further 42 days. His presence on the station meant that NASA could not carry out a long-duration mission of their own at the same time, as that would mean there were four permanent residents on the station, and the Soyuz lifeboat only carries three.

Therefore when Atlantis undocked on the 4 July, Dunbar was still on board along with the Mir 18 crew, including Norm Thagard. The old Mir crew rode back to Earth in new prone seats on the mid-deck. This basically involved the crew lying on their backs on the mid-deck floor with their feet in the storage lockers in front of them; it was felt that this was a better way for the long-duration crews to return to Earth. Nevertheless, Thagard broke the medical rules for his flight by walking out of the shuttle to the waiting astrovan.

Atlantis' mission to Mir had been made possible, as mentioned previously, by the Russian docking adapter. However, the corresponding adapter on the Mir was attached to the end of the Kristall module. In order for the shuttle to dock there without coming too close to the solar arrays, Kristall had to be moved from its normal position—on the docking node at right angles to the Mir base block—to the end of the Mir base block's docking node. In order to get around this necessity for future missions the next shuttle would bring an extended docking port which would be attached to the end of Kristall in its normal position. So it was that the shuttle Atlantis flying mission STS-74 arrived at the station in November 1995 with the new docking module in its payload bay. Two extra solar panels for Mir were transported affixed to the sides of the docking module, and were to be fitted to Mir at a later date by the resident crew. The Atlantis crew delivered many items to Mir, again far more

Shuttle docking adapter installed in Atlantis

than could be achieved by a Progress, and in fact more than the previous Atlantis mission, due to not having to carry Spacelab into orbit. The cargo consisted of food, water, replacement lithium hydroxide canisters, many items for future NASA research, plus many personal items for the Mir crew including a guitar that was put to good use by Canada's Chris Hadfield in a song describing Miss Dolly Parton!

NASA now had the opportunity to fly a further six astronauts on long-duration flights aboard the Mir station. Perhaps surprisingly there was not a great rush of volunteers to fill these positions, and NASA struggled to find 12 astronauts (including back-ups) who were willing to undergo the training in Russia for a year or more, and that fulfilled the criteria that the Russians had laid down for Mir crewmembers. The initial schedule for NASA's missions to Mir or increments as they liked to call them looked like this.

NASA increment	Prime	Back-up	Duration
2	Shannon Lucid	John Blaha	5 months
3	Jerry Linenger	Scott Parazynski	4 months
4	John Blaha	Wendy Lawrence	6 months
5	Scott Parazynski	Wendy Lawrence	4 months

Norm Thagard had two different back-ups during his training cycle for the first NASA increment—Bill Readdy and Bonnie Dunbar. Bill Readdy was apparently persuaded to be Thagard's non-flying back-up on the understanding that he would fly a later mission to Mir in which he would be launched by shuttle, but return to Earth in a Soyuz, something that no NASA astronaut had done up to that point. Ultimately Readdy was convinced to take up the position of Director of Operations (DOR) at Star City and later he commanded the mission that retrieved Shannon Lucid from Mir and delivered John Blaha.

Scott Parazynski was the first astronaut to fall foul of the Russian system. He had been training as back-up to John Blaha for the third U.S. increment aboard Mir when it was found that he was fractionally too tall for the existing Soyuz capsule. A new design of Soyuz was in the pipeline that would allow crewmembers of his stature, but this would not arrive soon enough, and he was removed from the program. Wendy Lawrence was doubly unlucky. She was initially removed from the program because she was too short for the Soyuz. And although she was reinstated in the program to succeed Michael Foale on the sixth increment to Mir, it was decided after the catastrophic events during Foale's flight that a crewmember capable of carrying out an EVA was required, and she proved to be too short to wear a Russian Orlan spacesuit, plus she had never been EVA trained at NASA, so she was removed from the crew rotation again to be replaced by David Wolf. This meant an accelerated training program for Wolf, as he had never served as a back-up crewmember, but he made the best of the situation and crammed his training in before launch. James Voss came into the breach as non-flying back-up for two of the increments, but he had already been assigned to an early ISS crew, and would make good use of his Mir training experience. After these changes the flight schedule now looked like this.

NASA increment	Prime	Back-up	Duration/EVA
2	Shannon Lucid	John Blaha	6 months
3	John Blaha	Jerry Linenger	4 months
4	Jerry Linenger	Michael Foale	$4\frac{1}{2}$ months + EVA
5	Michael Foale	James Voss	5 months + EVA
6	David Wolf	Andy Thomas	4 months + EVA
7	Andy Thomas	James Voss	$4\frac{1}{2}$ months

The remaining seven shuttle missions to Mir all followed the same pattern. The shuttle would bring the next replacement NASA crewmember and/or return with the old one. The shuttle would be fitted with a SpaceHab module in its payload bay to transport more supplies than previous shuttle flights to Mir, including experiments, food, clothing, water, and bring back experiment results and obsolete equipment, thus alleviating Mir's clutter problem slightly.

Shannon Lucid was the next willing volunteer for a long-duration mission to Mir. She was delivered aboard the shuttle Atlantis on mission STS-76 arriving on 24 March

STS-76 crew portrait, Lucid middle back row

1996, and was scheduled to stay aboard until Atlantis returned to collect her in early August.

A veteran of four previous shuttle missions, Lucid had eagerly volunteered for her mission to Mir, and undertaken the training in Russia with great zeal, seeing the chance to live on board a Russian space station with two Russians as a unique opportunity. Just over a month into her mission the final module, Priroda, to be added to the Mir complex, arrived. The contents of Priroda had been provided by many nations including Russia, America, Germany, France, and Canada, and truly reflected Mir's increasing role as a melting pot of international collaboration. Lucid's time on board Mir seemed to run more smoothly than had Thagard's before her. She was free to work at her own pace as she worked her way through the four-day task list that was updated by her NASA colleagues on the ground every day, she was also able to send and receive e-mails from friends and colleagues which helped to ward off any feelings of isolation. In mid-July Shannon was told that her mission would have to be extended to mid-September due to problems with the solid rocket boosters (SRBs) that had been stacked for Atlantis' mission to retrieve her (STS-79). A previous shuttle mission, STS-78, had experienced a problem with erosion of the field joints in its SRBs, which was a problem not dissimilar to that suffered by Challenger during its fateful flight in January 1986. It was thought that this erosion had been caused by a change in the type of adhesive used during assembly of the SRBs, and the boosters for STS-79 had been assembled in the same way. It was therefore decided to replace those boosters with the ones that had been set aside for STS-80, which used the original type of adhesive.

When the crew of STS-79 did dock with Mir on 19 September, Shannon had already broken two space records. On 7 September she had broken Elena Konda-kova's female duration record of 169 days, and on 17 September she broke the visitors (i.e. non-Russian) record of 179 days, which had recently been set by Thomas Reiter. The change of crew between Lucid and new arrival, John Blaha, was done in much the same way as the Russian crewmembers. They exchanged the seat liners for the Soyuz lifeboat capsule, and Lucid briefed Blaha as to the status and location of her many experiments and offered tips on living aboard the station. When Atlantis landed at the Kennedy Space Center Lucid had also set a new American duration record of 188 days. She adapted to Earth gravity more rapidly than expected, walking off the shuttle to the crew transport vehicle before later meeting with President Clinton.

Within NASA and the Phase 1 program there seemed to be two distinct camps of opinion on the collaboration with the Russians. Some felt that it was a business deal, pure and simple: NASA paid the Russians, plus offered the occasional seat on the shuttle, and in return the Russians provided training and room and board on the Mir station. Others believed that it was a proper partnership, or that at least it ought to be. With proper give-and-take on both sides, decisions being made jointly, and perhaps most importantly lessons learned on both sides, including language and working skills, organization of long-duration flights, and technology transfer between the two nations. This segregation within the program seemed to extend to the astronauts in training for the upcoming missions to Mir. Shannon Lucid for example, was definitely a member of the latter group; she saw the whole joint program as a massive opportunity, both for NASA and for her personally. She threw herself into her training, taking care to learn the Russian language and customs, and making sure that she integrated well with her assigned crew. This attitude served her well when time came to fly her mission, especially when it was extended. It was less clear which group John Blaha fell into; he was the only pilot-astronaut assigned to a Mir mission, and his background in the U.S. Air Force could certainly have given him reason to harbour a certain amount of distrust towards his Russian colleagues. Certainly his attitude cannot have been helped when the crew that he been training with—commander Gennadi Manakov and flight engineer Pavel Vinogradov—were removed from the flight just one month before his launch to Mir owing to a problem with Manakov's EKG. The pair were grounded and replaced with their back-ups, commander Valery Korzun and flight engineer Aleksandr Kaleri. Blaha did not even know who Kaleri was, but the insertion of Korzun worried him. He had carried out his winter survival training with Korzun and Michael Foale, and found the man to be condescending in the extreme; he could not imagine what spending four months under his command would be like. Blaha's troubles had not started there; for months he had struggled with the lack of support from the Phase 1 office at NASA. He had also become embroiled in an argument with both sides about carrying his own personal set of notes to Mir. His lack of expertise with the Russian language only added fuel to the fire, and all in all his training cycle had been a very difficult one. When launch day finally arrived for Blaha and the crew of STS-79, he was already exhausted.

Unfortunately, things did not get any better when he arrived at the station. Korzun and Kaleri welcomed him warmly, which made him feel a little better, but when he tried his first science experiments, with the shuttle still docked at the station, he immediately hit problems. As many astronauts and cosmonauts before him had discovered, it takes a lot longer to do even the simplest things in space, than it does on the ground; despite what it might say in the checklist. His first experiment was supposed to take only $1\frac{1}{2}$ hours to complete, start to finish; it ended up taking him 5 hours just to find all of the components in the sprawling cluttered station and put them together. In short, the reality was nothing like the organized straightforward training on the ground; things were not where they were supposed to be, and even if they were, there was no guarantee that they would be in working order. For a goal driven achiever, this was simply unacceptable, and the ground could not, and it seemed to Blaha, would not, help him. The ground support team acted as if Blaha was flying a shuttle mission, where everything is carefully cataloged and in its proper place; they did not seem to understand how it could be possible to not find something in a closed vehicle. Because of this lack of understanding, Blaha began to work longer and longer hours in his efforts to catch up with the timeline, which remained rigid and unaltered despite his pleas for it to be relaxed. Now, this may all sound very familiar in the light of the problems faced by the third crew aboard Skylab (they are in fact exactly the same problems) but nearly 30 years down the line, and on board a Russian station instead of an American one, with the additional issue of the language barrier only making the situation worse. It seems amazing to consider that NASA appeared

John Blaha portrait

to have learnt nothing, but it seems no-one from the Phase 1 program had paid any attention to the lessons of Skylab, and worse still, even when Phase 1 did learn from its mistakes, the newly learned lessons were not passed on, or not listened to, by the Phase 2 people.

Blaha soldiered on for the remainder of his four-month stay, and found, just as the Skylab 3 crew did, that things did improve over time. He became more adept at finding things on the cluttered Mir station, and the ground controllers eventually learned to relax their grip on the flight plan and leave some of the planning to Blaha himself. Nevertheless, he was quite relieved to hand over the reins to his replacement Jerry Linenger, who arrived aboard the space shuttle Atlantis on mission STS-81. Linenger was different from the previous Mir residents from NASA in that he had

Linenger before launch

only flown one shuttle mission, and he had been assigned hastily to comply with the Russian requirement that all astronauts for Mir be flight experienced. The plans for his increment were slightly different too; he was to carry out an EVA, the first by an American in a Russian Orlan spacesuit, and he was to spend the longest time on the station at that time. His attitude towards his Russian hosts was noticeably different from his predecessors too. He seemed to feel that NASA was paying the Russian Space Agency for a service, one that they were only barely providing. He seemed to have little interest in integrating himself into his two different Mir flight crews he would serve with. His main priority was to carry out his mission, and everything else

Soyuz-TM 25 crew, Tsibliyev, Ewald, and Lazutkin

was secondary (at least) to that goal. This attitude seemed likely to put him on a collision course with his Russian crewmates, and so it proved to be.

The Mir resident crew when Linenger arrived were still Valeri Korzun and Aleksandr Kaleri, and they welcomed Linenger just as warmly as they had John Blaha. As Linenger settled down to life on Mir, and began his schedule of experiments, Korzun and Kaleri noticed that he kept to himself and did not often join them for meals or other "social" occasions, but they let him be as the end of their increment slowly approached. In March 1997, a Soyuz arrived with the replacement Mir crew; it also carried a cosmonaut researcher, Reinhold Ewald from Germany, who would stay on board the station during the handover period.

Mir became a little crowded and cosy with six men on board, especially at meal times when all six would float around the table in the base block. The extra three-man crew on Mir required that the oxygen supply be supplemented by use of solid fuel oxygen generators (SFOG) into which tanks containing a chemical which produced oxygen when it was heated were inserted. These tanks were commonly referred to as "candles" by the cosmonauts. One evening as most of the crew gathered in the base block for the final meal of the day, Sasha Lazutkin, one of the new arrivals broke from the meal and went into the Kvant module to carry out the final candle burn of the day. Almost immediately he realized something had gone wrong. The candle started its burn in the normal way, but then he heard the tank hiss and, almost unbelievably, it burst into flames. Lazutkin was momentarily frozen by the sight before him, and even when he tried to shout a warning he was not heard. It was Ewald who saw the fire and screamed "Fire" in Russian. The rest of the crew finally realized what was happening, and Korzun dived into the Kvant module with Lazutkin. However, they quickly realized that it was not going to be easy to put out a fire in weightlessness that was being fed by the chemical reaction taking place in the candle. Fire extinguishers proved to be almost useless, and the fire continued to burn despite the efforts of the crew until slowly the solid fuel was consumed. By this time, the base block was almost filled with smoke, and the crew all donned oxygen masks; except Jerry Linenger, who was not in the base block. The fire alarm finally went off alerting Linenger in Spektr, he rushed into the node between Spektr and the base block and tried to find an oxygen mask for himself, the first mask he tried failed to work, the second was more successful. It was realized later that several of the emergency oxygen masks tried by all members of the crew were faulty. In addition, the first fire extinguisher that Linenger tried in Priroda was securely fastened to the wall and would not come off. The new commander, Vasily Tsibliyev, also tried to take an extinguisher from Priroda, but it too was securely fastened to the wall. It turned out that the transport straps put in place for Priroda's launch were never removed once the module reached orbit over a year and half previously. Emergency evacuation procedures called for the crew to prepare the two docked Soyuz spacecraft for departure, but one of the ships was on the other side of the fire, docked at the end of Kvant; this was the ship reserved for Korzun, Kaleri, and Ewald. Clearly, there was no way for anyone to reach it until the fire was put out. By the time the fire was finally out, most of the modules of Mir were filled with dense smoke and steam, and as thoughts turned to the effects of smoke inhalation Jerry Linenger reverted to his profession as a medical doctor. Of primary concern were the chemicals that made up the contents of the candle, and the residue of those chemicals in the smoky air. When the oxygen masks ran out, the crew donned surgical masks in an attempt to filter out any contaminants. The smoke slowly cleared, and the crew did the best they could to clean up the interior of the station, after which they washed and changed into clean clothes. Linenger carried out a health check on all of the crew checking their lungs for the effects of smoke, none of them appeared affected. After reporting the fire to the ground, the crew attempted to get some sleep.

Frank Culbertson was in the middle of a deep sleep when he was woken by his telephone. The call was to tell him about the fire, but it was not a Russian voice on the other end of the line, it was the voice of one of his support crew working at the Russian control center. Nobody from the Russian space program had thought to notify the head of the U.S. side of Shuttle–Mir that 12 hours ago one of their astronauts had just lived through the worst fire in spaceflight history.

Safety in space has always been the primary concern of NASA and the Russian space program. Both agencies had faced emergencies during their years of manned space flight, and sadly, both had suffered fatalities. The central tenet of the agreement between the two agencies was that each was responsible for the safety of the others' astronaut or cosmonaut crews. Missions on the shuttle and on Mir had taken place so far without incident, but the fire on Mir changed the perception of safety, especially in the minds of NASA, and the U.S. politicians and public. Suddenly, the perception was that Mir was risky and unsafe, and the Russian controllers were maverick and uncaring risk-takers. It did not help that many of NASA's own engineers felt that there was nothing to be learned from the Russians that they did not already know. Bearing in mind that Phase 1 was supposed to be the beginning of a long-standing partnership with the Russians, and that many of the lessons learnt here should bear fruit during the construction of the ISS, NASA took virtually no notice of anything the Russians did until something went wrong. Nothing exemplified the difference between the NASA and Russian way of doing things more than the attitude toward the fire. The Russians really thought that it was no big deal; they had fires on previous space stations, and there had been no problem putting them out and carrying on as normal. NASA, in contrast, spent large sums of money ensuring that every precaution against fire was taken; wiring, spacesuits, and non-flammable clothing were all checked and double-checked before flights. NASA's attitude was perhaps understandable given the fate of the crew of Apollo 1, who died in their spacecraft on the ground, but the Russians too had lost a cosmonaut in a fire on the ground and they did not believe such precautions to be necessary. The official Russian press release only intensified the distrust between the two parties; it stated that a "microfire", more likely described as a "small fire", occurred on the station for no more that 90 seconds, and that the crew easily extinguished it. Later, when both crews were back on the ground, this would be a major point of contention, Linenger was certain it had lasted about 14 minutes; other members of the crew thought it might have been about 5 minutes, perhaps more. Certainly none of them agreed that it was only one and a half minutes. As the days passed, the reactions to the fire began to calm down, but the seeds of discontent had been sown, both on the ground and on the space station Mir. The crew on board Mir, however, had plenty to occupy them. The resident crew, Tsibliyev and Lazutkin, were preparing a test in which they were to manually dock a Progress freighter. In itself, this was not an unusual occurrence; many crews had used the ability to manually dock Soyuz spacecraft and on occasion had used a remote control system to dock Progress ships from short distances when problems had surfaced with the automatic docking system. In this case, however, the crew was to attempt to dock the Progress from a range of about 8 km from the station. Tsibliyev would sit at the TORU controls that had been assembled in the base block, where he would maneuvre the Progress using two control sticks, one controlling its orientation, and the other imparting thrust fore and aft, left, and right. In front of him was a small screen, which transmited a view from the front of the oncoming Progress along with some simple radar information—that was all Tsibliyev had to judge the approach of the 7-tonne spacecraft as it hurtled toward the space station. Why, you might ask, carry out such a test? What contingency does it prepare the crew and station for? The answer has nothing to do with emergency scenarios or improving station operations, it has everything to do with money, a resource that the Russian space agency was desperately short of. The automatic docking system that was on board all Soyuz and Progress spacecraft, and the Mir space station was called Kurs, and it was manufactured in the Ukraine. Whilst the Soviet Union existed this was of no concern. But, since the break-up, and the independence of the Ukraine, it was now of utmost concern. The Ukrainian's charge a large amount of money for this equipment, and it occurred to the Russians that maybe they could manage without it; after all, each Progress spacecraft never returned to Earth, it burned up in the atmosphere taking its expensive Kurs apparatus with it. Already the Soyuz were using Kurs that had flown and returned many times before, but the same could not be true for Progress. If they could find a way to allow the resident station crew to perform the rendezvous and docking manually it would mean they no longer needed to install Kurs in every spacecraft. The test started badly. Tsibliyev could see nothing on the monitor in front of him. He urged Linenger and Lazutkin to look out of every window to try and find the approaching Progress, but they could not see it. Finally, Lazutkin spotted the spacecraft as it emerged from behind the station's solar arrays; it was close enough for Lazutkin to see the details of the craft's antennas and arrays. The monitor in front of Tsibliyev finally came on and showed that Progress would miss the station by barely 200 m. The three members of the Mir crew were shaken and angry, the Progress had been out of control and could have ended up anywhere. The exact nature of the near miss, or even the reason for carrying out the test was never communicated to the U.S. side; neither did NASA officials ask about it. Eventually Russian ground controllers told the NASA engineers that they had simply decided not to continue with the docking due to some software problems. Linenger, surprisingly given his mistrust of the Russian system, said nothing in his communications with his ground-based team; he assumed that they must know what had happened and how close it had been, but they did not.

Work on Mir carried on as normal, although Tsibliyev's attitude had noticeably changed. Linenger carried on with his schedule of experiments, his frustration with NASA's minute-to-minute planning growing by the day. The various systems on board Mir were not co-operating either: the Elektron system that produced breathable oxygen shut down, requiring the crew to burn more candles, which they were reluctant to do. The gyrodynes on the station—big gyroscopes that allow the station to be orientated without using thrusters—were proving temperamental, as were various power systems. Suddenly it seemed that nothing was working properly. The relationship between Linenger and his Russian crewmates had deteriorated, particularly with Tsibliyev. As the malfunctions on the station grew, it seemed to the Russians that Linenger was doing nothing to help; he would continue his personal routine regardless of anything else that might be going on. Tsibliyev resented that attitude. Russian ground controllers were concerned about the spacewalk that Linenger and Tsibliyev are due to carry out together; they feared that the antagonistic relationship between them was not an ideal basis for carrying out such a task. The spacewalk went ahead as planned, but rumours were abound of a "fight" between the two men whilst outside the station, a rumour which both men later strenuously denied. Meanwhile, NASA was getting nervous about the condition of Mir, to the point that it was considering not flying the next astronaut, Mike Foale, to the station. However, the shaky partnership continued when Foale launched on board the space shuttle Atlantis and arrived at Mir to replace Linenger. Foale was confident that he could enjoy a far better relationship with the Russian crew than Linenger, and immediately settled into the station's routine.

Michael Foale's outlook on his mission, and the whole partnership with the Russians was very different from Jerry Linenger's. From the very beginning of his

STS-84 crew (Mike Foale front right)

training he had ensured that he did the best he could to integrate himself into the Russian culture. As he learnt Russian, a difficult language that all NASA Mir candidates found hard, he made sure that he learnt more than just the technical vocabulary that was necessary for his job. He made friends with Russian colleagues, invited them to his home, and they in turn welcomed him into theirs. His conversational Russian became excellent, and the Russians appreciated the extra effort he was making. At the same time, Foale realized that working with the Russians was never going to be easy; the culture and attitude was so much different from that of the U.S., that complete trust and co-operation was never going to happen, at least not for many years. As soon as he arrived on board Mir he was determined to integrate himself fully into the Russian crew; it would mean biting his lip occasionally, but he wanted to gain the confidence of Tsibliyev and Lazutkin, and be trusted by them. There was certainly plenty of work to do, Mir was still suffering from malfunctions and coolant leaks, and Tsibliyev and Lazutkin were as busy as ever trying to catch up with the growing list of problems. Foale helped where he could, but the Mir crew were beyond exhaustion from the stresses and strains of the previous months, and still the problems with Mir's hardware continued.

Unbelievably, the Russian controllers decided to repeat the Progress docking test that had almost ended in disaster more than three months previously. Just as unbelievably, NASA officials who were this time informed of the upcoming test, said nothing, and Foale was not informed about it either. The lines of communication between the partners of this new space enterprise were virtually non-existent, and even the little that was being communicated was not being understood. Only a few days before the test was due to take place did Foale begin to question his commander about it. Tsibliyev, perhaps understandably, was reluctant to go into details, but, when pressed, Tsibliyev explained more about what was planned, and what had happened the last time they had tried. The reason for the repeat of the test was that the Russian engineers thought they knew the reason why the TORU monitor had failed to show any display. They reasoned that the Kurs radar signals, which had been turned on during the first test, had somehow interfered with the monitor's signal. The solution was simple to them; turn off the radar signals, and try again. Now the image from the camera on the front of Progress, if it worked, would be Tsibliyev's only source of information as he attempted to dock the 7-tonne spacecraft with Mir.

At the start of the test, the Progress craft was 7 km away from Mir and Tsibliyev was required to bring the spacecraft to a point about 50 m away from the Kvant docking port; all of this to take place whilst Mir was out of contact with controllers on the ground. To begin with, Tsibliyev was happy, at least the monitor was working this time, but he found it hard to make out the station from the clouds of the Earth behind it. Once again, his crewmates had their faces pressed against the windows searching for the Progress cargo craft, but they saw nothing. Lazutkin was the man that eventually spotted the Progress; it was very close and this time it was heading straight for the station. Tsibliyev ordered Foale into the Soyuz evacuation spacecraft. Moments later the Progress hit the Mir space station, the master alarm rang through the station, Foale felt his ears pop and it was clear to him that the hull of the station has been breached. He dived toward the Soyuz and prepared it for immediate departure, but Tsibliyev and Lazutkin remained on the station. Lazutkin knew where Progress had hit the station; he saw it with his own eyes. The Spektr module was now leaking its precious atmosphere, and the only course of action was to seal it off from the rest of the station. Unfortunately, this was not a simple exercise. Spektr, like all of Mir's modules, had cables and tubes snaking through its open hatchway and these needed to be removed before the hatch could be closed. Some cables were easy to remove, but Lazutkin could not find the attachment points for others, so he cut through them with a knife until finally the hatchway was cable free. Meanwhile Tsibliyev had begun to "feed the leak" by opening canisters of oxygen that were stored in Kvant 2; this action would keep the air pressure at a survivable level, for now. Lazutkin then tried to pull the inner hatch of Spektr closed, but even with Foale's help they could not do it, the air rushed past them and out of the puncture in Spektr's hull, making pulling the hatch closed impossible. The only other option was to find the original "lid" hatch cover that was in place in the node before Spektr docked; fortunately, these covers were stored in the node, and Lazutkin quickly grabbed one from its storage place on the wall— it was immediately sucked into place by the same escaping air that hampered their earlier efforts. The station had been saved from the immediate threat of depressurization, and the crew were safe, but the drama was not yet over.

Frank Culbertson barely had time to reflect that early morning phone calls had become something of a feature of the Phase 1 program, as he answered yet another one. Once again, he was amazed that a NASA astronaut had survived a life threatening accident, and no one from the Russian space agency had contacted him.

The drama aboard Mir was far from over, Spektr was now sealed off, and there was no immediate danger to the crew, but the impact of the Progress freighter had imparted a rotation to the station that could not be corrected as the attitude control computer was offline due to a lack of power. As the station drifted, the solar arrays could not track the Sun and generate power, the station's batteries took over the load, but they only had a limited life, and they were draining fast. Before communication with the ground was restored all power was lost, the lights went out, the gyrodynes and air circulation systems stopped; they established radio contact with the ground but expected to lose that at any moment. The crew turned on the radio in the Soyuz, which would be their only means of communication with the ground. The problem was that with the station still rolling, and no power to any of the station's systems, they appeared to have no means of stopping the roll and realigning the solar arrays with the Sun. Foale suggested that they use the Soyuz thrusters to regain control of the station, Tsibliyev was not keen on that idea; he had been been taught to preserve the Soyuz and its fuel at all costs, but the ground controllers finally agreed that this was the only option. After several attempts, the Soyuz was able to stop the roll and stabilize the station. Fortunately, when the station stabilized, it happened to be pointing its arrays at the Sun, and the batteries began to charge, it was a slow process, but Mir finally cames back to life.

The Phase 1 program was at the end of its tether. The feelings of distrust and hopelessness overwhelmed Frank Culbertson and his team, pressure from U.S. politicians as well as from within NASA began to tell; the message seemed to be "we just can't trust the safety of our astronauts to the Russians". The first meaningful partnership between the space superpowers was at a crossroads, and most of NASA wanted to stop right there. In Russia, things were completely different; to the officials of their space program this was simply another bump in a long road. Salyut 7 was a good example of the kind of repairs that cosmonauts could accomplish. That station had literally been brought back from the dead, and it was obvious to them that they could do the same thing again. NASA continued to consider its options including having Foale return to Earth with Tsibliyev and Lazutkin on board the Soyuz instead of waiting for a shuttle to pick him up. The implications of their future actions were plain to see, if they removed Foale early, Phase 1 was over, and so too, almost certainly, was the future co-operation for the ISS. Foale's safety was obviously important, but it would be naive to think it was the only concern. For the first time, NASA was not in control of the destiny of one of its own; they had to trust completely the Russian space agency's ability to keep their man safe, and they were not sure that they did. As much as they tried to impress upon the Russian officials their concerns, and their opinions, the truth was that the Russians would continue to do things as they saw fit; it was, after all, their station.

Over the next few weeks, the station experienced several power drops, resulting in the station again drifting out of control, and the Soyuz had again to be used to regain the station's attitude. NASA officials were worried that a further shuttle docking would not be possible; if such a fault occurred during the last phases of docking, it would be disastrous. Russian officials, such as Valery Ryumin, were determined to press ahead with their schedule for the station. This included sending a guest cosmonaut, Leopold Eyharts from France, to the station with the next long-duration crew. NASA felt that would put an unnecessary drain on Mir's limited resources, and should be postponed, but Ryumin would hear nothing of it; the French had paid for their mission, and he saw no reason to cancel it. Part of the preparations for the French mission was the repair of Spektr, which would be carried out by means of an internal spacewalk by Tsibliyev and Lazutkin. They were to enter Spektr in spacesuits, which would be cramped at best, and find and repair the hole made by the Progress collision. It was important that power be restored from the Spektr solar arrays to run the French mission experiments.

The Russian Mir crew of Tsibliyev and Lazutkin meanwhile, felt sure that they would take the blame for the whole affair. Tsibliyev in particular, was certain that he would never be allowed to fly in space again, Lazutkin was less sure of that, but equally certain that their flight pay and bonuses would be affected, perhaps even lost entirely. With this pressure already on their shoulders, and the weariness they had borne with months of failures, repairs and the consequent lack of sleep, they did not feel up to the task of repairing Spektr, although they did not say so outright to the ground. To add insult to injury, Tsibliyev suffered from a heart arrhythmia during an exercise period designed to test his health for the upcoming repair. This effectively ruled him out of the work, and put Foale in the spotlight as the only other man who could join Lazutkin to carry out the repair. The final straw for Tsibliyev was when he was told that he could not participate in the spacewalk; he broke down. Lazutkin and Foale did what they could to console him, but his depression extended beyond their capacity to help. Later, Lazutkin was preparing cables for the upcoming repair work, and a badly written checklist caused him to disconnect the power to the main station computer; again the station lost attitude control, and tumbled. The recovery process was long and tiring for the crew, and now they really needed to come home; they had been through more than any crew in history, more even than the crew of Apollo 13. Finally realizing that the crew were at the end of their tether, Russian ground controllers reassigned the repair work to the next crew. Although somewhat disappointed, Tsibliyev and Lazutkin were also relieved, and the mood on the station lightened. Tsibliyev joked with the ground when asked about Foale's new haircut, "I told him I would cut his hair when the cargo ship comes; it came and he said, 'well, one has come and it hit us, so cut my hair'." They were ready to come home.

Michael Foale would not be coming home with them, but the question of who should succeed him, if anyone, was well underway. Wendy Lawrence was due to replace Foale with the launch of shuttle Atlantis on STS-86, but should her seat remain empty to allow Foale to return home, or should the program continue? It now seemed unlikely that NASA would pull out of Phase 1 entirely. In the event, Lawrence was not chosen to replace Foale. Much earlier in the program it had been identified that Lawrence was too short for the Russian EVA spacesuit; this had not been seen as a problem since no EVA was scheduled for her increment, but the problems on Mir had changed that. Now it was decided that each NASA crewmember had to be capable of carrying out an EVA if it became necessary, and Lawrence could not. She would have to be replaced. This was not to prove as easy as it may of sounded. NASA astronauts that wanted to be a part of the Shuttle–Mir program were extremely thin on the ground, especially after the recent events on Mir. In fact, the only man that they could find was David Wolf, who had previously flown on STS-58, a Spacelab life sciences flight lasting 14 days, which seemed perfect experience for a flight on board Mir. Unfortunately, Wolf's career had taken something of a down-turn since then. The occasional brush with the law, and his love of good living meant that he was not likely to be assigned to a shuttle flight for a long time, if ever. For Wolf the choice was straightforward, it was Mir or nothing. The choice was not a trivial one however. Having never served as a back-up to a Mir mission mean that Wolf was starting from scratch, and with much less time to train than all of the previous long-duration crewmembers. If Wolf was not ready in time, or simply could not fly because of illness or injury, Shuttle–Mir would probably be over, and Culbertson and the rest of NASA knew it.

Tsibliyev and Lazutkin's long mission neared its end on 8 August 1997 when Soyuz-TM 26 docked with Mir bringing with it the next expedition crew of Anatoli Solovyov and Pavel Vinogradov. The old crew packed the existing Soyuz with items to be returned to Earth, and prepared for re-entry. Even the landing of their Soyuz was not without incident. The rockets that were designed to fire moments before the capsule hit the ground, to soften the landing, failed to do so, and the crew landed hard, fortunately without injury to either of them. Tsibliyev knew that now the inquisition could begin.

On board Mir all eyes were now on the repair of Spektr. The plan called for the new crew to install a new hatch to replace the cover hurriedly put in place by Foale and Lazutkin. This new hatch incorporated electrical connectors to link up Spektr's solar arrays; the loss of the power from these arrays had significantly reduced Mir's overall power stores. Whilst the crew were inside Spektr, they would also try to find the puncture for future repair.

When the internal spacewalk did take place, it was mostly successful. Vinogradov initially had a problem with his suit, and when they did get inside Spektr it took longer than anticipated to connect up the cables to the new hatch, but eventually they managed to do it, and power flow from the solar arrays was restored. Unfortunately, the crew were not able to find the source of the leak from inside the module, and it was decided that Solovyov and Foale would later try to discover the puncture from outside. That EVA also failed to find the leak, and reluctantly, following further failed attempts to find the leak by the shuttle crews of STS-86 and STS-91, Spektr was abandoned.

STS-86 launched on 25 September 1997 and carried a crew of seven, including Wendy Lawrence and David Wolf; only Wolf would get to remain on the station. Foale was overjoyed to see his friends and colleagues after such a long mission, and they were relieved to see him safe and well.

Wolf's expedition to Mir seemed boring and incident free compared with both Linenger's and Foales, but he did carry out an EVA with Solovyov, and despite some initial difficulties with the strict Russian commander, he came to be respected by his Russian crewmates.

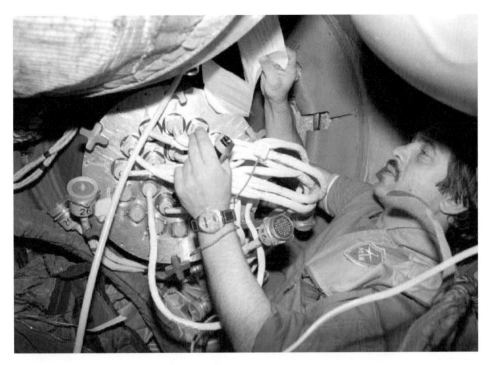

Vinogradov adjusts the hermaplate hatch leading to Spektr

Andy Thomas had never expected to fly to Mir, he had simply been Lawrence's and then Wolf's back-up; but when Wendy could not fly, he found himself in the last seat to the space station. He gladly accepted the opportunity, and launched on STS-89 Endeavour's first and only docking with the Mir space station. His flight increment was the smoothest of all of the Phase 1 missions; he got on well with his crew of Talgat Musabayev and Nikolai Budarin, and enjoyed the postponed visit of Frenchman Leopold Eyharts. As he says himself, "I think [my mission] was probably the most placid of all of them. The first person, Norm Thagard [had] ... a lot of problems to do with the fact that he was the first, and [I don't think he had] a lot of the things that you need to sustain yourself. So that must have made it tough for him. I don't think they had the email situation worked out ...

"Shannon [Lucid's] flight got extended—because of shuttle problems, actually. So she had to stay up there six months instead of four, and that would have been tough, I think. She has a very good spirit about it, though ...

"For Jerry [Linenger] there was the fire, of course, and for Mike [Foale] there was the depressurization. So they had some exciting times on theirs. [And] David [Wolf] had a number of power failures during his.

"Mine," Thomas concludes, "was fairly placid by comparison, which I think is testimony to the capability of the Russians to restore operations, to bring the system

Foale reunited with family after landing

back on line, which I think they did well, because I think they recognized . . . that they were on the world stage and needed to prove that they could do it, and they did that."

Space shuttle Discovery collected Thomas at the end of his increment, and marked the end of the Phase 1 program with the final shuttle docking to Mir. Valery Ryumin was a member of that final crew, his first space flight for 18 years; he reported back to Russia with first hand information on the current state of the Mir space station. Andy Thomas returned to Earth having completed the final Mir increment successfully, he now looked forward to flying to the ISS once it was built. Frank Culbertson too wanted to fly a long-duration mission to the ISS; he had certainly earned it after a long and hard experience as head of Phase 1, butting heads day to day with Ryumin and other Russian officials, as well as those within his own organization. For Tsibliyev and Lazutkin their concerns proved well founded: initially Tsibliyev was blamed completely for the collision, but Culbertson and others felt strongly that Russian ground controllers and trainers were at least as responsible. Eventually, the officials caved in and exonerated Tsibliyev of all blame, and both cosmonauts were paid their full flight bonuses. However, neither were to fly in space again. Tsibliyev was promoted to command the cosmonaut training division, and

STS-91 and Mir-25 in-flight crew portrait

NASA's Mir astronauts

Lazutkin was grounded for unknown medical reasons following several assignments to ISS back-up crews.

Shuttle–Mir proved to be a vital component of the future plans for the ISS. Vital too for the part it played in allowing two disparate nations to settle their differences, and solidify its synergies ahead of the greatest joint program of space history, the International Space Station.

11

The International Space Station ... at last

Looking at NASA's recent history of trying to build a space station, the casual viewer might think it a miracle that anything ever got off the ground, but this would be an unfair assessment. NASA did its best in an ever changing political world, both at home and abroad, and through several different presidencies and administrators. In some ways, the agency was its own worst enemy; its designs were always leading edge and therefore very expensive and time consuming to build, and therefore unlikely to be approved by Congress in times of financial constraint.

However, in 1998 the first hardware of the International Space Station (ISS) stood ready. After almost 14 years of political turmoil, numerous redesigns, and countless billions of dollars, NASA and its partners were finally ready to launch hardware into space. These first components were the Russian FGB, or Functional Cargo Block, called Zarya, which was a module that would give the early station attitude control, and the U.S. Node 1, or Unity, which was a connecting module to allow for further expansion of the station. Zarya, which was based on the TKS design from Chelomei's design bureau back in 1969, was launched in November 1998 on a Russian Proton rocket, and was followed by the space shuttle Endeavour on mission STS-88 in December. The crew connected the two modules together, and carried out spacewalks to electrically link the modules, Sergei Krikalev was a mission specialist on this crew, as well as being a member of the planned first expedition. STS-96 followed in May 1999 with supplies for the first expedition, and carried out further spacewalks to "get ahead" before the Russian Service Module was launched.

January 2000 saw a seemingly unrelated meeting of Congress that created the Iran Non-Proliferation Act. In short this act disallowed any U.S. companies or organizations (including government agencies) from paying money to any country helping Iran with its nuclear program. Russia is one such country. On the face of it, this did not seem of any concern to the ISS program which was still early in its construction, and early too in its partnership with Russia. NASA already had a contract with Russia to provide 11 Soyuz and Progress spacecraft for crew rotation

Proton launching Zarya

Expedition 1 crew

and cargo delivery to the ISS that would last until April 2006. This contract was exempt from the new act as it had been agreed before the act's creation. It surely would not become a problem later, because NASA planned to have its own Crew Return Vehicle in place by then, and the European Space Agency (ESA) was creating its own cargo delivery system (the Automated Transfer Vehicle) which was due to be ready before that date.

As the Service Modules' delays increased, it was decided to fly another shuttle logistics mission, STS-101, in May 2000 to bring further supplies. The much delayed Service Module, Zvezda, was launched in July 2000, and docked with the orbiting Unity/Zarya combination. A Progress craft M1-3 docked at Zvezda's rear port shortly after and replenished the fuel that had been used during the launch and docking procedures. The space shuttle Atlantis was launched in September on mission STS-106 to carry out the final outfitting of Zvezda prior to the arrival of Expedition 1; Atlantis also fired its engines to raise the orbit of the fledgling station. STS-92 launched in October with the first element of the truss assembly, and an additional docking port to be attached to the Unity module, and Discovery's engines further raised the ISS orbit before it left.

Finally, the station was ready to receive its first crew. Soyuz-TM 31 was launched on 31 October 2000 carrying Soyuz commander Yuri Gidzenko, Expedition 1 commander William Shepherd and flight engineer Sergei Krikalev.

They docked with the station on 2 November, and entered shortly after. They were restricted to the two Russian modules initially, as there was not enough power

Proton launching Zvezda

available to facilitate the use of Unity. The space shuttle Endeavour brought that additional power in the form of the first solar array truss, on the mission STS-97 in December 2000, and over the next few days the two huge solar arrays were deployed, and the internal and external connections made to enable the flow of power to the station.

The next major component of the ISS, the Destiny research laboratory, was launched aboard STS-98 on 7 February 2001 and connected to the station on the 10 February. Spacewalks were carried out to facilitate external connections between Destiny and Unity, and Atlantis raised the ISS orbit before leaving.

The Expedition 1 crew moved their Soyuz from the rear port of Zvezda to the downward or nadir (Earth-facing) port of Zarya, thus freeing the rear port for more Progress dockings. The first crew's stay was nearly at an end, and when Discovery was launched on STS-102 with the Expedition 2 crew aboard it was time for NASA's first crew rotation of the ISS program. The new crew consisted of commander Yuri Usachev, and flight engineers Susan Helms and James Voss. The rotation plan called for a rather complex one-at-a-time swap of crewmembers from ISS to the shuttle.

STS-98 crew in an unusual pose

This routine was dropped on later crew rotation missions. STS-102 landed on the 21 March ending the first expedition to the ISS.

In April 2001 Endeavour was launched on mission STS-100, bringing the Canadian robotic arm to the station. The Canadarm-2 was vital to future station construction and operations, so this was an important mission; all was not to go smoothly, however. On the night of 24 April one of the command and control computers on board the ISS went offline unexpectedly, and the crew was woken to troubleshoot the problem, causing NASA to extend Endeavour's mission by two days, and thereby conflicting with the upcoming first Soyuz "taxi" mission to replace the ISS Soyuz with a new one. The Russians agreed that the taxi Soyuz would be launched on time assuming that by the time it arrived at the station, the shuttle would be gone. If Endeavour was still there, as it was, then the Soyuz would remain in formation with the ISS. The Canadarm-2 passed its first tests successfully, but there were still lingering doubts over some of the computer systems on the ISS. However, Endeavour undocked and left the station on 29 April, leaving the way clear for

Expedition 3 crew

Soyuz-TM 32, or EP-1 as the Russians called the mission. This mission was already controversial because of the inclusion of Dennis Tito, the first space tourist, on the crew. NASA argued that he was unqualified, especially on NASA systems, and therefore a danger to the station; the Russians countered that he was bringing valuable revenue to their space program, and therefore indispensable. It was finally agreed that Tito would limit himself to the Russian modules unless escorted by a U.S. crewmember; in reality Tito was happy to listen to music and gaze out of the windows in Zvezda during his six days aboard the station, but he did help out by preparing meals and doing odd jobs for the resident crew. Problems with the Canadarm-2 persisted, and the launch of the next shuttle mission was postponed, as it would require the use of the arm to install the first dedicated airlock module, called Quest, on the station. STS-104 was launched with the Quest airlock in July, and it was successfully installed without any major problems.

STS-105 followed in August with the next crew rotation, Expedition 3, commander Frank Culbertson, Soyuz commander Vladimir Dezhurov, and flight engineer Mikhail Tyurin. After all of the trials of the Phase 1 program, Culbertson made it back into space, eight years after his commanded shuttle mission STS-51. This crew would have the distinction of being the only humans in orbit during the 9/11 terrorist attacks, and it soon emerged that Culbertson had gone to flight school with the pilot of the airliner, Chic Burlingame, that hit the Pentagon. He wrote a letter home about the attack that was published on the NASA ISS website, in which he concluded "What a terrible loss, but I'm sure Chic was fighting bravely to the end. And tears don't flow the same in space ..."

The next taxi mission, Soyuz-TM 33, arrived on the 23 October, carrying the crew of Viktor Afanasiev, flight engineer Konstantin Kazeev and French astronaut Claudie Haigneré. This crew spent eight days on board the ISS, where Haigneré performed experiments for ESA under the Andromeda program before the crew's return in the older Soyuz-TM 32 spacecraft that had been docked to the ISS since the end of April.

STS-108 carried the Expedition 4 crew of commander Yuri Onufrienko and flight engineers Dan Bursch and Carl Walz to the station on the 7 December, and left on the 15 December. This crew had a fairly quiet time, apart from the failure of a wrist joint on the stations robot arm, and had not received any visitors until Atlantis arrived with a new segment for the truss element of the station in April 2002.

The third taxi mission, Soyuz-TM 34, docked on the 27 April, with the crew of Russian commander Yuri Gidzenko, and Italian researcher for ESA, Roberto Vittori. It also included the second space tourist in its crew. This time there was no controversy. South African Mark Shuttleworth had been completely trained on all aspects of station operations, both Russian and U.S., and had a scientific program to work through. This mission was to be the last launch of a Soyuz-TM spacecraft; the next launch would be the newer Soyuz-TMA version. Of course, the older Soyuz-TM 34 would remain docked to the ISS as the emergency crew return vehicle until it was replaced by the newer version. Again, the taxi crew spent eight days on board the station before leaving in the older Soyuz-TM 33 spacecraft on the 5 April.

In 2002 NASA canceled both the CRV (Crew Return Vehicle) and the ISS Habitation module. At a stroke this immediately reduced the maximum crew of the ISS to three, as this is obviously the number of crew that a Soyuz can carry. Without the Habitation module there would be nowhere for additional crew members to eat and sleep. By doing this, NASA broke several agreements with its International partners and created a great deal of ill feeling within the scientific community as it was fully realized that it took at least two crewmembers full-time just to keep the station running, this leaving barely one crewmember to carry out any science at all. Certainly it would not be possible for any international astronauts to now fly to the station to tend their own experiments, as there would be no room for them to live aboard, the station. The obvious choice in the short term would be to make use of more Russian Soyuz and Progress spacecraft, but this was not possible. All of a sudden the Iran Non-proliferation Act which had seemed unimportant a year ago was thrown into sharp relief. NASA could not buy anymore of anything from Russia without breaking the law. However, the end of that contract was still five years away, and the shuttle would be able to pick up any slack until then, wouldn't it? In the meantime, NASA planned to go ahead with development of the Orbital Space Plane (OSP) concept, although this would not be ready until at least 2008.

I'm sure you are beginning to see a pattern forming by now. Every NASA plan would be first put on hold, and then canceled, to be replaced by something even more expensive, and even further into the reaches of time, after they had already spent way over their budget on the first project. The OSP project was canceled in 2004, and replaced by the new Crew Exploration Vehicle (CEV), of which, more later.

On 5 June 2002, STS-111 was launched with the Expedition 5 crew of Valery Korzun, Peggy Whitson, and Sergei Treschev. Whitson was flying in the newly created role of NASA Science Officer. This was seen as something of an appeasement to the science community, who argued that not enough science was being carried out on the ISS. When Endeavour landed ending the Expedition 4 mission, Bursch and Walz had set a new U.S. duration record of 195 days. During the mission of Expedition 5, the space shuttle fleet was temporarily grounded by the discovery of cracks in the hydrogen lines for the main engines. The problem was fixable, but delayed all subsequent shuttle missions. Therefore, it was not until October that Atlantis was launched on STS-112 with more components for the truss segment, plus supplies for the resident crew.

At the beginning of November, the next taxi mission, Soyuz-TMA 1, docked with the station, originally the EP-4 crew was to have included American singer Lance Bass, but not all of the money for his flight was forthcoming, and the Russians were strict. He was replaced by Russian cosmonaut Yuri Lonchakov. The rest of the crew consisted of commander Sergei Zalyotin, and ESA astronaut Frank De Winne of Belgium who would carry out a program of experiments for ESA. The taxi crew returned to Earth in Soyuz-TM 34, leaving the new Soyuz-TMA 1 for the resident crew.

STS-113 arrived at the station at the end of November 2002 with another truss segment, and the Expedition 6 crew. The new ISS crew included commander Ken Bowersox and flight engineers Don Pettit and Nikolai Budarin, with Pettit also taking on the role of NASA Science Officer. Budarin had flown twice before to the Mir space station, including a launch on the U.S. shuttle Atlantis on STS-71. Bowersox had flown on the U.S. shuttle four times previously; twice as pilot, and twice as commander. Pettit was making his first space flight on Expedition 6, replacing Don Thomas who had to stand down for medical reasons. The crew of STS-113 had also undergone a crew change when the original pilot Gus Loria was replaced by Paul Lockhart, who had just flown on STS-111, when Loria injured his back at home which caused him to miss too much training time. Both of these changes required new mission patches for both STS-113 and Expedition 6, which made the original STS-113 patch with Loria and Thomas on it a rare item.

On the 1 February 2003 NASA's worst nightmare scenario came to pass. The space shuttle Columbia was destroyed during re-entry, after a non-ISS scientific mission. The reasons for Columbia's tragic loss have been well documented elsewhere and need not be repeated. However, the loss would have massive implications for the immediate future of the ISS, its future beyond April 2006, and indeed the future of NASA itself.

The big date that NASA had been aiming at for the last couple of years was 19 February 2004. This was the date that the station would be triumphantly proclaimed to be "U.S. Core Complete", which was to say that certain key U.S. modules (including the important Node 2) would be connected to the station, and NASA could progress to fulfilling its promises to its international partners to launch their modules to the station. ESA and Japan in particular were keen to get their Columbus and Kibo (respectively) science modules launched and docked to the station after many delays.

Clearly, immediate changes to operations would need to be made, starting with the very next expedition. With Expedition 6 only just settled on the station, any changes were not urgent in nature, apart from the question of how best to get that crew home. Originally, they had been scheduled to be swapped with the Expedition 7 crew at that time consisting of commander Yuri Malenchenko, Sergei Moschenko and NASA's Edward Lu, who would fly up on shuttle mission STS-114. However, changes to the personnel of Expedition 7 were taking place even before the Columbia accident. Sergei Moschenko was replaced by Aleksandr Kaleri, apparently because Moschenko's English was not up to scratch.

On 1 April 2003, exactly two months after the Columbia accident, it was announced by NASA that the Expedition 7 crew would be reduced from three to two, and would consist of commander Malenchenko and NASA Science Officer and flight engineer Lu. Kaleri was bumped to back-up the flight of Expedition 7, and would later fly on board Expedition 8 with NASA's Michael Foale in command. The reduction of crew from three to two would help to reduce the demands on the station's food and water supply which would now only be replenished by Progress freighters, which could not carry anything like as much cargo as the shuttle. The crew rotations would now be carried out using the Soyuz-TMA spacecraft, which meant that the taxi missions were on hold for now, blocking a source of income for the Russians.

Therefore, on 26 April 2003 Soyuz-TMA 2 was launched with the two-man Expedition 7 crew of Malenchenko and Lu, replacing Bowersox, Pettit, and Budarin, who would return to Earth aboard Soyuz-TMA 1. Their return was not entirely straightforward, as the crew landed 460 km short of their target due to a computer error which commanded the capsule into a ballistic re-entry path, subjecting the crew to higher g loads than normal.

Soyuz-TMA 3 launched with the Expedition 8 crew of Foale and Kaleri in October 2003, with the Expedition 7 crew coming home in the older Soyuz-TMA 2. Foale now commanded a space station after his eventful flight aboard Mir. This expedition was trouble-free by comparison. This process was repeated in April 2004 with the launch of Soyuz-TMA 4 with the Expedition 9 crew of Gennady Padalka and Michael Fincke. The Expedition 10 crew of Leroy Chiao and Salizhan Sharipov were launched to the station on board Soyuz-TMA 5 on 14 October 2004. The pair stayed on board the ISS for 192 days and landed back on Earth on 24 April 2005 after enduring many ongoing problems with the station's Elektron oxygen generating system. Sharipov attempted repairs for many days but the system was still offline and awaited the efforts of the Expedition 11 crew of Sergei Krikalev and John Philips.

On 26 July 2005 the space shuttle returned to flight after two and half years on the ground. The flight of STS-114 Discovery was a life saver to the ISS program as it relieved so many problems at once. The replacement CMG that had failed so long ago would be repaired, and the gathering rubbish and clutter on board the station would be greatly reduced by the Raffaelo module due to be delivered by Discovery which would also bring desperately needed new supplies including food and water. Following missions would restart the construction process. Unfortunately, that all fell flat after about a minute and a half of flight, when it became apparent that the fixes to the external tank (ET) of the shuttle were not all that NASA had hoped they would be. Newly installed cameras showed a large piece of foam from the ET falling away as the solid rocket boosters (SRB) separated. The chunk of foam missed Discovery, but it was quickly realized that it could have inflicted just as much damage as that visited upon Columbia two and a half years earlier. Although Discovery continued with her mission, the shuttle was once again grounded until further notice pending the solution to the foam problem. When Discovery reached the ISS it carried out a pre-planned pitch over maneuvre so that the crew on board the ISS could photograph Discovery's underside to check for missing and/or damaged tiles. Unfortunately, many damaged tiles were discovered, seeming to prove beyond doubt that the external tank foam problem was just as bad as ever, and that re-grounding the shuttle was the right thing to do. The next shuttle mission, STS-121, was postponed until at least May 2006, after the deadline for NASA's existing agreement with Russia for Soyuz had run out.

The Expedition 12 crew of Bill McArthur and Valery Tokarev were launched to the station on 1 October 2005 on board Soyuz-TMA 7 along with space tourist Greg Olsen. The new residents were relieved by Expedition 13 in March 2006. On 11 October 2005, Soyuz TMA-6 landed with the crew of Expedition 11, and space tourist Greg Olsen. Sergei Krikalev and John Philips had spent 179 days in space, and Krikalev was now the most travelled cosmonaut or astronaut in history with a grand total of 803 days in orbit from his two missions to Mir, two space shuttle flights, and two stays aboard the ISS.

On 26 October 2005, the House of Representatives came to NASA's rescue when it voted to allow the space agency exemption from the Iran Non-Proliferation Act. This meant that they could buy Soyuz and Progress spacecraft from Russia until 2012. This allowed NASA to concentrate on finishing construction of the station without having to deliver new crews on the shuttle as well, and assured U.S. astronauts access to the ISS. All future crew rotations were to be carried out by the Soyuz craft, and cargo delivered by the Progress, with much heavier items being lifted into orbit by the shuttle once it finally returned to flight. Certainly this would prove to be much more cost effective for NASA as each Soyuz flight costs in the region of only $65 million, instead of at least $500 million for each shuttle launch.

The date 1 November 2005 marked an important day in the history of the ISS. The station had been continually occupied for 1,826 days, or 5 years. Since the first crew's arrival, the ISS had grown considerably and now weighed 183 tonnes, with a habitable volume of $424 \, m^3$; by comparison Skylab weighed 90 tonnes with a volume of $361 \, m^3$, and Russia's Mir weighed 110 tonnes with a volume of $380 \, m^3$. There had been 97 visitors on board the station from 10 countries; and 29 had lived aboard as

Expedition 13 patch without and with Reiter's name and German flag

members of the 12 station expedition crews. Russian cosmonaut Sergei Krikalev was the only one so far to have served as a member of two resident crews.

On the 7 December 2005 the crew of Expedition 13 was announced, and consisted of station commander Pavel Vinogradov and U.S. astronaut Jeffrey Williams, they were to be joined on Soyuz TMA-8 by Brazilian Marco Pontes, who had trained at NASA as a mission specialist for a time until the agreement with Brazil for ISS components had been discontinued. He was to stay about one week.

It was not until July 2006 that STS-121 finally got off the ground, almost a year since the less-successful-than-hoped STS-114. This mission had originally been scheduled as a somewhat anti-climatic follow up to STS-114; it would simply repeat and revalidate most of the feats of the previous mission. It had not existed in the original pre-Columbia flight schedule, and had been added to reinforce the fact that the shuttle was once again safe to fly, hence its "out of sequence" numbering. However, after the failure of STS-114 to completely validate the new foam application process on the ET, the mission took on a new importance, rather than being just a follow-up, it would be an important milestone in proving the shuttle to be fit for purpose. It would also deliver a new crewmember to the ISS, bringing the permanent crew up to a full strength of three for the first time since Expedition 6. Thomas Reiter of the ESA would also be the first long-time crewmember from outside the U.S. or Russia.

The foam failures on STS-114 had come from an area of the tank that contained the attachment points for the SRBs, specifically, from an aerodynamic ramp in front of the attachments. NASA engineers had come to the conclusion that the best solution would be to remove those ramps completely, and remove the risk of foam shedding at the same time. However, these ramps had been previously thought to be essential aerodynamic aids for launch, and some were nervous about what effect their removal might have. On 4 July 2006, all worries were laid to rest when Discovery launched flawlessly into orbit with its crew of six, plus ISS crewmember Reiter. This time the ET performed perfectly, and no damage was seen on Discovery's thermal protection system. The crew carried out the same checks as the flight of STS-114, and the same pitch-over maneuvre that was to be a part of all ISS docking approaches. Once docked at the station, the crew delivered more supplies, carried out two space-walks to test future safety options, and made a few repairs on the ISS. The crew undocked, and landed on 17 July after leaving Thomas Reiter on the ISS to officially join the Expedition 13 crew.

The Expedition 14 crew consisted of U.S. commander Michael Lopez-Alegria, and two flight engineers, cosmonaut Mikhail Tyurin and astronaut Sunita Williams. The first two crew were launched on board Soyuz-TMA 9, whilst Williams flew to the ISS aboard STS-116 in December 2006. This continued the trend set by Thomas Reiter flying to the ISS aboard STS-121, the intention being to carry on the practice of sending the third crewman to the ISS aboard the shuttle to allow the Russians to sell the third seat on the Soyuz to potential researchers or tourists. In fact the format of the expedition crews was to change somewhat from Expedition 14 onwards. The crew size remained at three, but the third crewmember would change more frequently so that each expedition would have two or three different flight engineers, some from the partner nations such Japan, Canada, and ESA.

With the return of STS-121, NASA had renewed confidence in the shuttle's abilities, and wanted to press on with the construction of the ISS. STS-115 with its crew of six was to add two pairs of new solar arrays to the ISS to provide power for the future Columbus and Kibo laboratories, and they would be supplemented by the crew of STS-117 in 2007. The space shuttle Atlantis was ready for its first space flight in four years, although not without some difficulty. A hurricane was due to hit the Kennedy Space Center whilst Atlantis was sitting on the pad, and the decision was taken to roll Atlantis back to the Vehicle Assembly Building (VAB), however, just as the shuttle was halfway back, the forecast changed and NASA managers decided to take Atlantis back to the pad after all. This was the first time in the shuttle's history that such a move had been made. On 9 September 2006, Atlantis hurled herself into orbit to begin a 12-day construction mission. After the now standard checks of the shuttle tile system, which showed no damage at all, Atlantis docked with the ISS and prepared for the first of three spacewalks that would be carried out by this crew. First the new solar arrays, and the truss they are attached to, had to be lifted out the shuttle's cargo bay using the shuttle's robot arm and handed over to the station's robot arm before they were attached to the station, this involved great precision and communication between Dan Burbank and Chris Ferguson, who were operating the shuttle's arm, and Canadian Steve MacLean who was on board the ISS operating its

Canadarm-2. The new truss was left in the grasp of the stations arm until the next day, when spacewalkers Joe Tanner and Heidemarie Stefanyshyn-Piper removed the launch restraining bolts and began wiring the new arrays to the station after the truss was firmly attached to the existing structure. Steve MacLean and Dan Burbank continued the work on the flight's second spacewalk, and the following Thursday the commands were sent to begin the unfurling of the new solar arrays. There were some concerns at this stage that the arrays might stick (they had, after all, been packed in their storage box for over three years due to the delays in the shuttle program) but they deployed perfectly to their full length of 240 ft. A third spacewalk by Tanner and Stefanyshyn-Piper completed the work by removing the last restraints that allowed a radiator to unfurl, and also upgraded the stations communications system. STS-115 had been an outstanding success, but it was not without its little last minute drama. Two days after undocking from the ISS, as the crew prepared for re-entry, a small object was observed by ground radar floating just below the shuttle's belly. They immediately postponed the planned de-orbit burn to give them time to diagnose this discovery. The next day permission for re-entry was given after the crew had carried out another check of the thermal protection system with the shuttle arm and boom; it was later decided that the object was almost certainly a plastic tile gap filer, and that the other smaller bits of debris probably came from the cargo bay, and were a result of Atlantis' prolonged down time between flights.

One more flight remained in 2006; the December launch of STS-116 with the shuttle Discovery. The crew of seven contained five rookies, amongst them Sunita Williams, who was to become the second new member of the Expedition 14 crew on the ISS. She would remain on board for the next six months, crossing over the period when Expedition 15 took over command of the station. ISS resident Thomas Reiter would be brought back to Earth after his long stay on the ISS that began with the launch of STS-121. There were more criteria for Discovery's launch than previous missions, mostly due to the time of year; NASA did not want the shuttle to still be in orbit during the New Year crossover, as it was felt that the on board computers may have a hard time dealing with the new date. A second consideration were the orbital thermal conditions during December which made it preferable to launch between 7 and 26 December, because the shuttle must launch as the ISS passes into the correct orbital plane for the shuttle's ascent path, this dictated a night-time launch, the first since the Columbia accident. Night launches had been ruled out by the safety commission due to the needs of ground-based photography to inspect the shuttles tiles and ET during launch; however, NASA felt that the last two missions had provided enough confidence over the launch performance of the redesigned ET to allow this requirement to be waived. Lastly, an engine firing by the docked Progress freighter was required to boost the ISS orbit enough to allow a docking on flight day three of the mission regardless of the day of launch. The first attempt at this firing had failed, possibly due to the station's unbalanced current configuration, but a second attempt, carried out a week later with a software patch to compensate for the station's off-axis center of mass, was successful and allowed Discovery's countdown to begin as scheduled. Unfortunately, the first launch attempt was scrubbed due to unacceptable weather at the launch site. The second try was scheduled two days later, as the weather forecast for a 24-hour turnround were thought unlikely to be any better. On Saturday 9 December, mission STS-116 got underway despite the weather looking to be uncooperative for most of the countdown. The spectacular light show put on by the shuttle's main engines and SRBs was enough for some of the launch photography to be carried out, but in fact such pictures were not really necessary, as the launch proceeded flawlessly, and as Shuttle Program Manager Wayne Hale said, "We're not relying on those ascent-based observations for the safety of that particular flight, we're relying on the inspection of the heat shield, which we do in excruciating detail on orbit now to make sure they're safe to come back." On flight day three, as scheduled, Discovery docked with the ISS and immediately the crew set to work, Sunita Williams installed her seat liner in the docked Soyuz for use in an emergency evacuation of the station, and at that moment became an official member of the Expedition 14 crew. Thomas Reiter, had until that moment, been part of the resident crew, but now joined the STS-116 crew in preparation for his return to Earth. Rookie mission specialist Nicholas Patrick, born in England, hoisted the new solar array truss out of the payload bay and handed it over to the ISS robot arm operated by Sunita Williams, where it would wait until the next day and the first of the planned Extravehicular Activities (EVAs). The first EVA was to be carried out by veteran space walker Robert Curbeam and his partner Rookie Swedish astronaut Christer Fuglesang. They successfully installed the new segment of truss on the end of it's existing length. The first EVA was so trouble-free that they were able to tackle a couple of tasks from the next EVAs schedule in order to ensure that the mission kept ahead of the timeline in case something unexpected happened later in the mission. Something unexpected duly did occur the very next day, when ground controllers attempted to fold up one of the existing solar arrays to allow them to be moved during a later mission, and also to allow enough clearance for the adjacent arrays installed during STS-115 to begin to rotate to track the Sun. The array refused to fold up by the required amount, getting stuck after only about half of its length had been retracted. Ground controllers decided to press ahead with the rest of the mission's EVAs whilst they worked on possible solutions. Curbeam and Fuglesang carried out a second EVA to rewire the station's power supply in readiness for more solar arrays; so successful were their efforts that the spacewalk ended an hour early. The third EVA, which was to feature ISS resident Sunita Williams' first EVA with Curbeam, continued with the rewiring efforts and also reconfigured some of the station's cooling systems as well. Ground controllers had decided to add a task to this EVA, if there was time, to get the spacewalkers to look at the stuck solar array and try to provide more information on the problem, and if time allowed, carry out repair efforts. The spacewalking crew managed to coax the array further into its box, but a stubborn part on one side of the array stopped further progress as the astronauts' time for this spacewalk had run out. Ground controllers decided to add a fourth EVA to the already crowded flight plan for the very next day, that they hoped would finally put the array back in its box. However, the extra time needed for an additional EVA did not come without a price. The crew would have to sacrifice a planned contingency landing day, which would mean that Discovery would have less time to play with in the event of unacceptable weather at the primary landing sites at the Kennedy Space

Center (KSC), Edwards Air Force Base, or Nothrup Strip at White Sands in New Mexico. The White Sands landing site has only ever been used once in the shuttle's history, at the end of Columbia's third test flight; engineers on the ground say that it took years to get all of the gypsum out of Columbia's nooks and crannies. The fourth EVA proved to be successful when Curbeam managed to coax the last of the array into its box. This prepared the way for the crew of STS-120 which will fly toward the end of 2007, to relocate the array to the other end of the station. Discovery undocked from the ISS the next day and began preparations for its return to Earth. Commander Mark Polansky said, "It's always a goal to try and leave some place in better shape than it was when you came and I think we've accomplished that due to everyone's hard work. And so with that, I hope we're really on our way to a great start for assembly completion." Discovery and her crew attempted to land at the first opportunity, but were waived off when the weather at both KSC and Edwards became unacceptable; they would have to wait another orbit (an hour and a half) before they could try again. Then, with minutes to go, KSC's weather decided to co-operate, just, and the crew were given a "go" to carry out the de-orbit burn, Discovery completed the mission after nearly 13 days with a smooth touchdown on KSC's concrete runway ending STS-116 and the long-duration mission of some six months for Thomas Reiter.

The year 2006 had been a successful one for NASA and its ISS partners. STS-116 had added more power to the station, and the stage was set for the next shuttle flight, STS-117, to finish that work. Atlantis was due for launch on 15 March after being rolled out to the pad on 15 February and everything appeared to be on schedule to meet that launch date, when on 26 February a freak hailstorm tore through the Kennedy Space Center causing damage to the exposed nose of the external tank. At first it was hoped that the damage would not be too bad, and that repairs could be carried out at the pad, but inspection revealed at least 1,000 points of damage, additional damage was found on Atlantis' left wing and a rollback to the Vehicle Assembly Building (VAB) was essential. When the shuttle got back to the VAB and close inspections were possible it was finally determined that there were 2,644 points of damage. The repair effort was clearly going to take some time, and the launch schedule for the year was now under threat. The original launch schedule had called for STS-117/Atlantis to be launched mid-March, followed by STS-118/Endeavour in June, and STS-120/Atlantis in September, with STS-122/Discovery rounding out the year in November. The turnaround of Atlantis after STS-117 for launch again on STS-120 would also have been very difficult in such a short timescale. On the 16 April, NASA announced a new flight schedule which still allowed four missions to the ISS in 2007. The complete schedule can be found in more detail in Appendix A, but the summary is:

STS-117/Atlantis (8 June 2007)
STS-118/Endeavour (9 August 2007)
STS-120/Discovery (20 October 2007)
STS-122/Atlantis (6 December)
STS-123/Endeavour (14 February 2008)

STS-124/Discovery (24 April 2008)
STS-119/Endeavour (10 July 2008)

NASA was not having a very good start to year in other respects as well. A great deal of unwanted media interest was focused on the space agency when astronaut Lisa Nowak, who had flown on STS-121, was arrested on 5 February. It was alleged that Nowak, a U.S. Navy Captain, had tried to kidnap her perceived rival for the attentions of fellow astronaut, U.S. Navy Commander Bill Oefelein. Oefelein has just recently returned from his first spaceflight as pilot on STS-116 in December last year. The court case is not due to take place for some months but NASA relieved Nowak of her duties at the Johnson Space Center, and returned her to the Navy, stating that the Navy were better equipped to deal with this case than NASA.

Better news came on 7 April, when the Expedition 15 crew of Fyodor Yurchikhin and Oleg Kotov, along with space tourist Charles Simonyi were launched successfully to the ISS. Simonyi is the fifth space tourist to fly by virtue of an agreement with Space Adventures Ltd. Simonyi is best known as being the man behind software applications such as Word and Excel. Commander Fyodor Yurchikhin and flight engineer Oleg Kotov will replace Expedition 14 commander Michael Lopez-Alegria and flight engineer Mikhail Tyurin, the third ISS crewmember Sunita Williams will remain on board until she is replaced by Clay Anderson. On 2 April, Lopez-Alegria became the longest flying U.S. astronaut when he broke the existing record set by Expedition 4 of 196 days, however, this record may be broken by Sunita Williams if she still returns to Earth on board STS-118/Endeavour which is now delayed until August 2007. After the change in the space shuttle's launch schedule, some thought was being given within NASA to returning Williams on STS-117 in June rather than waiting for STS-118. At the time of writing no final decision had been reached. Even if she does not break the endurance record, Sunita has already broken a record of a different kind when she took part in the Boston marathon on 16 April. She was an official participant running on the stations treadmill whilst fellow astronaut Karen Nyberg took part on the ground. Running on the treadmill is an important part of any astronauts exercise regime whilst on a long-duration mission, and Williams, who is an accomplished marathon runner, had been training for this run for most of her flight so far. She finished the "course" in 4 hours, 23 minutes, and 10 seconds.

In the remainder of 2007 it is important to NASA's schedule to complete the ISS, with several missions adding essential parts to the station. STS-117 will add more solar arrays—extra power that is essential to the new modules that are due to be delivered over the next two years. STS-118 will add another segment of the station's truss structure. In addition it will see the flight of Barbara Morgan, Christa McAuliffe's back-up for the tragic STS-51L Teacher in Space mission. STS-120 adds another vital component to the growth of the ISS when it launches with Node 2, now called Harmony. Harmony will serve the same purpose as Unity currently does, as an interconnecting module for future labs such as Columbus and Kibo. STS-122 will take the long awaited European Space Agency lab, Columbus, to the station, before

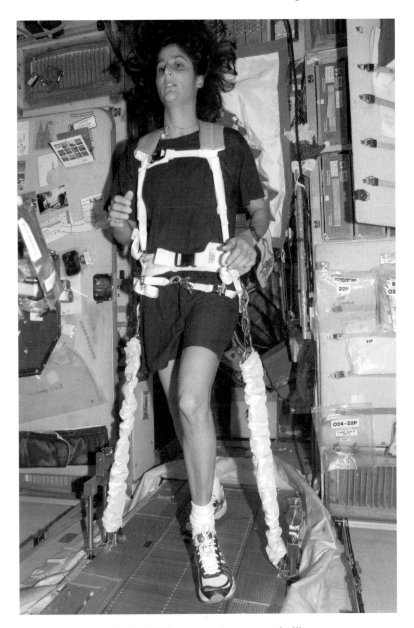

Sunita Williams running on treadmill

2008 when STS-123 and STS-124 will attach the various parts of Japan's Kibo lab to the station. Beyond that will be the installation of Node 3, more solar arrays and truss segments, before the ISS is declared complete (see color plates for the complete ISS assembly sequence).

12

The future for manned space stations

NASA was in a depressed state in 2003. Many within the agency had to consider their role in the loss of the Space Shuttle Columbia, and many others continued to mourn the loss of the seven astronauts. More worrying still was the fact that management shortcomings, which had been a contributing factor to the loss of Challenger, seemed to have returned and played a part in another tragedy. The U.S. public again questioned the need for NASA and space exploration, and even within the agency itself the loss of direction and purpose had instilled itself in the minds of the staff at all of the NASA's centers. The White House had felt for some time that a new injection of energy and exploration was needed, not just for the employees of NASA, but to a public that felt America was losing its way, bogged down by conflict, both at home and abroad.

When the final report from the Columbia Accident Investigation Board was delivered in August 2003 it made a great many specific recommendations for the safe return to flight of the space shuttle program. It also suggested changes and improvements that were not specifically required for flight, but that were felt to be necessary for NASA's future. It made the point firstly that "One is the lack, over the past three decades, of any national mandate providing NASA a compelling mission requiring human presence in space," and secondly that "Since the 1970s, NASA has not been charged with carrying out a similar high-priority mission that would justify the expenditure of resources on a scale equivalent to those allocated for Project Apollo. The result is the agency has found it necessary to gain the support of diverse constituencies. NASA has had to participate in the give and take of the normal political process in order to obtain the resources needed to carry out its programs. NASA has usually failed to receive budgetary support consistent with its ambitions. The result, as noted throughout Part Two of the report, is an organization straining to do too much with too little." In the previous chapters we have seen that all too frequently NASA had not been given the funding or ongoing support it needed to see programs to their conclusion, the CAIB recognized this and hoped that the U.S. Government would do something about it.

The CAIB report also highlighted the fact that the space shuttle's days were numbered, the loss of two shuttles was clearly too many, and most now accepted that the shuttle was an inherently dangerous design. The almost complete lack of a crew escape system, plus the sheer technical complexity of the space shuttle meant that it was not acceptable to continue flying astronauts on it. Most people accept that space travel is always going to be a dangerous occupation, but why make it more dangerous than it needs to be? The public perception of NASA putting their space crews at risk, apparently without any concern, could not continue, a new spacecraft would be needed to carry future crews to Earth orbit and beyond, one that encompassed more safety features, and viable escape options when things go wrong.

In the early part of 2004, The White House took two steps to improve NASA's future, and give the agency a sense of direction once again. First, President Bush announced in a speech at NASA's headquarters on 14 January, the steps required for humans to return to the Moon, and eventually to land on Mars. Several key milestones relating to the space shuttle, the ISS, and future plans for the Moon and Mars were revealed along with the need to create a new space vehicle.

The Space Shuttle and International Space Station
- Complete assembly of the International Space Station, including the U.S. components that support U.S. space exploration goals and those provided by foreign partners by 2010.
- Return the space shuttle to flight as soon as practical, based on the recommendations of the Columbia Accident Investigation Board.
- Retirement of the space shuttle by the end of 2010

The Moon
- Undertake lunar exploration activities to enable sustained human and robotic exploration of Mars and more distant destinations in the Solar System.
- Starting no later than 2008, initiate a series of robotic missions to the Moon to prepare for and support future human exploration activities.
- Conduct the first extended human expedition to the lunar surface as early as 2015, but no later than the year 2020.

Mars
- Conduct robotic exploration of Mars to search for evidence of life, to understand the history of the Solar System, and to prepare for future human exploration.
- Conduct robotic exploration across the Solar System for scientific purposes and to support human exploration. In particular, explore Jupiter's moons, asteroids, and other bodies to search for evidence of life, to understand the history of the Solar System, and to search for resources.
- Conduct advanced telescope searches for Earth-like planets and habitable environments around other stars.
- Develop and demonstrate power generation, propulsion, life support, and other

key capabilities required to support more distant, more capable, and/or longer duration human and robotic exploration of Mars and other destinations.

- Conduct human expeditions to Mars after acquiring adequate knowledge about the planet using robotic missions and after successfully demonstrating sustained human exploration missions to the Moon.

New spacecraft
- Develop a new crew exploration vehicle to provide crew transportation for missions beyond low Earth orbit.
  - Conduct the initial test flight before the end of this decade in order to provide an operational capability to support human exploration missions no later than 2014.
- Separate to the maximum practical extent crew from cargo transportation to the International Space Station and for launching exploration missions beyond low Earth orbit
  - Acquire cargo transportation as soon as practical and affordable to support missions to and from the International Space Station.
  - Acquire crew transportation to and from the International Space Station, as required, after the space shuttle is retired from service.

The second step came two weeks later when an Executive Order formed a commission comprising several industry leaders. It would be their job to outline the best way for NASA to achieve the goals that President Bush had set in his speech; this commission had only four months to report its findings back to the White House.

At last it seemed that NASA would have a clear path forward that it had been craving since Apollo 17 left the surface of the moon so many years ago. It was clearly understood what was required of the space agency, for the first time in many years they would be striving for achievable goals, rather than pushing frontiers that it couldn't hope to reach. Most importantly, the funding for these goals had been committed, and as long as NASA made good progress, the money would continue to flow. It remains to be seen how this policy will be carried over from one President to the next, but for now NASA has a clear goal to work towards.

Perhaps the largest single task ahead of NASA is the development of a replacement spacecraft for the space shuttle, the Crew Exploration Vehicle (CEV), or Orion as it would later become officially known. What would such a vehicle look like? Would it be wings and wheels again, like the space shuttle, or would a simpler design be a surer bet for success? A significant complicating factor was the need for such a spacecraft to not only fly to Earth orbit and the ISS, but also be adaptable enough to form the basis of a Moon and Mars orbiter. In September 2004 NASA issued contracts to eight aerospace contractors to begin studies into the kind of designs which would fullfil the following requirements:

- Support a minimum crew of four (NASA preferred six) from the Earth's surface through mission completion on the Earth's surface.

- Have a mass less than 15–18 tonnes (the precise value to be determined in preliminary contract studies).
- Have an abort capability during all phases of flight. Preferably such abort capability would be available continuously and independent of Launch Vehicle (LV) or Earth Departure Stage (EDS) flight control.
- Integrate with the Constellation Launch Vehicle (LV) to achieve low Earth orbit.
- Integrate with the Earth Departure Stage (EDS) to achieve lunar orbit.
- Integrate with the Lunar Surface Access Module (LSAM) to achieve lunar surface mission objectives. Preferably the CEV would be capable of transferring consumables to and from the EDS and the LSAM.

Perhaps not surprisingly the selected companies came back with very different design solutions, they did, however, agree on some basic principles. Namely that it would be most cost effective to make use of either an existing launch vehicle, or one derived from existing technology. This launch vehicle would also make use of extra stages or strap-on boosters to make launches to the Moon or Mars possible from the same core rocket. They also agreed that a four-man craft, at least for Earth orbit missions, would be ideal, and should weigh less than nine tonnes. By June 2005, NASA had narrowed the contractor list down to two; Lockheed Martin, and a joint team of Northrop–Grumman and Boeing, these two "finalists" would build a CEV of NASA's design, and the decision between the two would be made without either party having to build a prototype. NASA's own design had changed somewhat from the original requirements, the crew had grown from four to six, and the launch weight had grown with it, to 30 tonnes. The increased weight also rather narrowed down the list of launch vehicles available, in fact no existing rocket was considered suitable to launch the CEV in its new form. A new launcher, derived from existing shuttle technology would have to be created. In fact NASA seemed intent on pushing its own design for both the CEV and launch vehicle rather than embracing the designs submitted by experienced aerospace contractors after months of detailed technical and practical study. It seemed clear that NASA had never intended to make use of the innovative designs that many of the contractors had come up, and had always planned to make use of its design. Many industry experts felt that NASA's basic design assumptions were flawed, and likened the situation to the initial designs of the Apollo spacecraft that took Americans to the moon nearly forty years earlier.

Whatever NASA's intentions the winning contractor was announced on the 31 August 2006, it would be Lockheed Martin that would build the new spacecraft named Orion. The spacecraft's new name had been officially announced the previous day, but unfortunately some of that fire had been stolen when astronaut Jeff Williams, speaking during a press conference on board the ISS, had let the name slip eight days earlier.

With that announcement made, NASA's attention turned to the launch vehicles that would be used for Orion. The first, most basic, is called Ares 1, it is otherwise known as "the stick". This is the launch vehicle that will be used for all of Orion's Earth orbit missions, including those that rendezvous with the ISS. Initially it appeared as if Orion's great weight would be far too much for the shuttle solid rocket motor derived Ares 1, many within the industry feel that NASA has more problems with its design than it is letting on, but during recent press conferences NASA has assured everyone that Ares 1 will be ready on time for Orion's first flight, not thought to take place sometime in 2012.

On 13 December 2004 Sean O'Keefe resigned as Administrator of NASA. He had maintained this position for three years, through the Columbia disaster and the troubled planning for the shuttle's return to flight.

The space shuttle returned to flight with Discovery flying STS-114 in July 2005. This flight was not without its problems, but NASA is now back in the business of flying space shuttles and completing the construction of the International Space Station (ISS).

The new NASA Administrator, Mike Griffin, has vowed to reverse the fortunes of a beleaguered agency, and focus on Project Constellation. On 28 September 2005 Griffin said that the shuttle and ISS, indeed the whole of the U.S. manned space program for the past three decades, had been mistakes! He said NASA lost its way in the 1970s, when the agency ended the Apollo program of moon visits in favor of developing the shuttle and space station, which can only orbit Earth. These decisions can be directly connected to the Apollo mode decision made during the 1960s.

"It is now commonly accepted that was not the right path," Griffin said. "We are now trying to change the path while doing as little damage as we can. It cannot be done instantaneously."

Only now is the nation's space program getting back on track, Griffin said a week after the announcement that NASA aims to send astronauts back to the moon in 2018 in a spacecraft that would look like the Apollo capsule and would be carried into space by a rocket built from shuttle components.

When asked whether the shuttle had been a mistake, Griffin said, "My opinion is that it was. It was a design which was extremely aggressive and just barely possible, especially with the amount of funding allocated to the problem." He added on the subject of the ISS which was started in 1999, "Had the decision been mine, we would not have built the space station we're building in the orbit we're building it in."

Griffin's statements have sparked a great deal of analysis of the space shuttle and ISS programs. Hindsight of course is a wonderful thing, but at the time, at the end of Apollo and Skylab, NASA had very little choice about its next manned spaceflight program. Had money been no object, then clearly things would have been different, but Congress and President Nixon would only allow a certain amount to be spent; there were, as always, other priorities. Had NASA pushed for more moon flights, or missions to Mars they would simply have been turned down, and possibly left with no manned program of any kind. NASA hoped that by going ahead with the shuttle, compromised design though it was, they would eventually be able to add the other components, such as the space station, at later dates. To a certain extent this turned out to be the case, but it took far longer than NASA had envisaged, and it had already cost the lives of the seven Challenger shuttle astronauts before anything else was built or flown. The birth of the ISS has already been covered in earlier chapters,

Shenzhou 5 crew—Liwei Yang

but one thing is worth considering. Griffin suggests at the end of his statement that he would not have built the station in its current orbit. He is presumably alluding to the fact that NASA agreed to change the inclination of the ISS orbit to 51 degrees in order to enable the Russians to launch payloads and crews from their launch site at Baikonur. The detrimental effect of this decision was that the shuttle effectively had its payload to ISS orbit capability cut by as much as 30%; or to put it another way, the change added a further 10–15 shuttle flights to the building schedule. Griffin apparently views this concession as a mistake, but imagine if the ISS had been placed in its original 28-degree orbit, out of the reach of the Russians, the station would have had to be abandoned after the Columbia accident in February 2003, as NASA would have had no other means to reach it.

The Chinese, on the other hand, have made their intentions quite clear. They now have two successful manned flights under their belts with the launches of Shenzhou 5 in October 2003, and the two-man launch of Shenzhou 6 in October 2005.

Shenzou 6 crew—Junlon Fei and Haisheng Nie

The second of these flights is more significant because it lasted for just over five days, had a crew of two, and for the first time for the Chinese that crew carried out scientific experiments. To call the Shenzhou 6 spacecraft a "mini space station" would be taking things too far. After all, Shenzhou is an evolution of the Russian Soyuz spacecraft, although it is larger and has been changed a great deal from the original Soyuz design. However, the Chinese clearly have that kind of development in mind with this design. The orbital module is larger than Soyuz, and has its own propulsion and solar arrays that allow for autonomous flight. This means that the orbital module can be left flying, and carrying out automatic experiments after the crew has left the module and returned to Earth. Just such a mission profile was followed during the first manned flight of Shenzhou in 2003, the orbital module remained circling the Earth fulfilling a six month long military imaging mission. It also means that these modules could be launched to attach to an existing space station by themselves allowing the station to grow. It is thought that the orbital module comes in different sizes for different mission profiles.

They will have a manned station in orbit by 2015, and whilst nobody could say that this is an accelerated program, the station will be of their own design, and not borrowed Soviet/Russian technology. The Chinese are in space to stay, and seem keen not to repeat the unfocused programs of NASA and its partners, but to take one step at a time in a logical fashion. In more recent times (11 January 2007), the Chinese have angered the rest of the world with their testing of an ASAT (anti-satellite) resulting in the destruction of an obsolete weather satellite; such tests have not been carried out by either the U.S. Air Force or the Soviets for about twenty years. This further underlines the fact that the Chinese are following their own agenda both in space and on the ground, and have little regard for the opinions of the rest of the world.

The Russians too have plans for the future despite their more limited financial resources. A design intended to replace the venerable Soyuz and Progress spacecraft is on the drawing boards, and it is called Kliper. Design work on this new spacecraft began back in 2000, but its configuration has changed many times since to reflect both the needs of the Russian, and ISS space programs, and of course the budgets of the agencies involved. The specification is now set to carry six people to and from Earth orbit, plus carry 500 kg of cargo/supplies. The new spacecraft will have a service life of 10 years or 25 flights. The design was first revealed to the public in February 2004 at a press conference held by Yuri Koptev of RKK Energia; however, by April 2005 no funding from the Russian government had been forthcoming according to Valery Ryumin of RKK Energia. Good news for Kliper came in June 2005, when the European Space Agency (ESA) seemed to commit themselves to the development of the project. This would allow Kliper to be launched from the ESA Korou launch site as well as the existing Russian facilities. The support from ESA could mean that Kliper should launch sometime in 2011. Kliper will be launched by a Soyuz-3 booster, and in August 2005 a model of the Soyuz-3 booster with Kliper atop was shown at the Moscow Air and Space Show, MAKS-2005. Japan has also shown interest in the project as involvement would give them independent access to the ISS and its own Kibo Spacelab without requiring seats on the U.S. space shuttle. However, in the summer of 2006, ESA changed its plans, and forced RKK Energia to revisit the design of the Kliper spacecraft, it now seems unlikely that the Kliper will ever fly as Russia's focus had returned to the Soyuz, and a possible upgrade of that spacecraft.

Space stations have literally come a long way since 1971 and the launch of Salyut 1. The Soviets/Russians have arguably made the greatest leaps, both in terms of hardware design and crew organization and motivation. NASA, however, has learned to apply its greater levels of technology relevantly and with great effect. It has, perhaps, taken them longer to embrace the finer points of crew interaction and scheduling, probably understandable all the time they were flying the space shuttle as well. Now NASA, with Orion, has the opportunity to make great strides beyond low-Earth orbit, to the moon and Mars, but what of the future of the ISS, this is much less clear. Orion will service the ISS once the shuttle completes construction of the station before the end of 2010, and Russia will continue to send Soyuz spacecraft and Progress cargo ships, the ESA will also send its ATV to the ISS for replenishment of consumables for the crew. Beyond that, Russia has plans to expand the station with

Orion approaches the ISS

more modules, but that initiative is solely reliant on finding the money to finance it. The ESA and Japan will at long last have their labs to carry out research in, and that will probably keep both organizations busy for some time to come. Beyond all of these possibilities, the future of the ISS is unknown, in fact none of the participating nations are saying very much about the future; presumably they are all too tied up in getting construction completed.

The next space stations will probably not even be in Earth orbit, stations orbiting the moon and Mars seem more likely to be the next stage of development, and clearly this represents an even greater challenge, for both man and technology.

ISS completed

Appendix A

Flight schedule and International Space Station crewing

Scheduled launch date	Flight designation	Flight crew	Mission objectives	Mission insignia
9 April 2007	Soyuz-TMA 10	Commander: Fyodor Yurchikhin Flight engineer: Oleg Kotov Flight engineer: Simonyi	Deliver Expedition 15 crew to ISS	
Resident ISS crew	Expedition 15	Commander: Fyodor Yurchikhin Flight engineer: Oleg Kotov Flight engineer: Sunita Williams		
8 June 2007	STS-117 Atlantis	Commander: Frederick Sturckow Pilot: Lee Archambault MS1: Steven Swanson MS2: James Reilly MS3: Patrick Forrester MS4: John Olivas	ISS-13A: 2nd starboard truss segment, install S3/S4 solar arrays	

| 9 August 2007 | STS-118 Endeavour | Commander: Scott Kelly
Pilot: Charles Hobaugh
MS1: Rick Mastracchio
MS2: Tracy Caldwell
MS3: Dafydd Williams (Canada)
MS4: Barbara Morgan
MS5: Clayton Anderson (Exp 15—Up)
MS5: Sunita Williams (Exp 14—Down) | ISS-13A.1:
SpaceHab-SM, 3rd starboard truss segment. Deliver third ISS crewmember (Anderson) | |

| Resident ISS crew | Expedition 15 | Commander: Fyodor Yurchikhin
Flight engineer: Oleg Kotov
Flight engineer: Clayton Anderson | | |

| 6 October 2007 | Soyuz-TMA 11 | Commander: Yuri Malenchenko
Flight engineer: Peggy Whitson (ISS Cdr)
Flight engineer: Sheik Muszaphar Shukor | Deliver Expedition 16 crew to ISS | |

| 20 October 2007 | STS-120 Discovery | Commander: Pamela Melroy
Pilot: George Zamka
MS1: Stephanie Wilson
MS2: Scott Parazynski
MS3: Paolo Nespoli
MS4: Douglas Wheelock
MS5: Daniel Tani (Exp 15—Up)
MS5: Clayton Anderson (Exp 15—Down) | ISS 10A:
Node 2, sidewall | |

Scheduled launch date	Flight designation	Flight crew	Mission objectives	Mission insignia
Resident ISS crew	Expedition 16	Commander: Peggy Whitson Flight engineer: Yuri Malenchenko Flight engineer: Daniel Tani		
6 December 2007	STS-122 Atlantis	Commander: Stephen Frick Pilot: Alan Poindexter MS1: Hans Schlegel MS2: Stan Love MS3: Rex Walheim MS4: Leland Melvin MS5: Daniel Tani (Exp 16—Down) MS5: Leopold Eyharts (Exp 16—Up)	ISS 1 E Columbus	
Resident ISS crew	Expedition 16	Commander: Peggy Whitson Flight engineer: Yuri Malenchenko Flight engineer: Leopold Eyharts (ESA)		

14 February 2008	STS-123 Endeavour	Commander: Dominic Gorie Pilot: Gregory Johnson MS1: Takao Doi (Japan) MS2: Richard Linnehan MS3: Michael Foreman MS4: Robert Behnken MS5: Garrett Reisman (Exp 16—Up) MS5: Leopold Eyharts (Exp 16—Down)	ISS 1J/A JEM ELM PS / SLP-D1	
Resident ISS crew	Expedition 16	Commander: Peggy Whitson Flight engineer: Yuri Malenchenko Flight engineer: Garrett Reisman		
8 April 2008	Soyuz TMA-12	Commander: Sergei Volkov Flight engineer: Oleg Kononenko Flight engineer: Ko San or Yi So-yeon	Deliver Expedition 17 crew to ISS	

Scheduled launch date	Flight designation	Flight crew	Mission objectives	Mission insignia
Resident ISS crew	Expedition 17	Commander: Peggy Whitson Flight engineer: Sergei Volkov Flight engineer: Oleg Kononenko		
24 April 2008	STS-124 Discovery	Commander: Mark Kelly Pilot: Kenneth Ham MS1: Michael Fossum MS2: Karen Nyberg MS3: Ronald Garan MS4: Stephen Bowen MS5: Akihiko Hoshide	ISS 1J JEM PM, RMS	
10 July 2008	STS-119 Endeavour	Commander: Pilot: MS1: Michael Gernhardt MS5: Sandra Magnus (Exp 17—Up) MS5: Peggy Whitson (Exp 17—Down)	ISS 15A: 4th starboard truss segment, S6 array	
11 September 2008	STS-125 Discovery	Commander: Scott Altman Pilot: Gregory Johnson MS1: John Grunsfield MS2: Michael Massimino MS3: Andrew Feustel MS4: Michael Good MS5: Megan McArthur	HST service 4 Non-ISS mission	

Resident ISS crew	Expedition 17	Commander: Sergei Volkov Flight engineer: Sandra Magnus Flight engineer: Oleg Kononenko	
9 October 2008	STS-126 Atlantis	Commander: Pilot: MS1: MS5: Koichi Wakata (Exp 17—Up) MS5: Sandra Magnus (Exp 17—Down) ISS ULF-2	
Resident ISS crew	Expedition 17	Commander: Sergei Volkov Flight engineer: Oleg Kononenko Flight engineer: Koichi Wakata	
8 September 2008	Soyuz TMA-13	Commander: Salizhan Sharipov Flight engineer: Michael Fincke Flight engineer: Deliver Expedition 18 crew to ISS	

Appendix B

Mission patches

Mission patches have been part of manned space flight for such a long time that it is easy to forget their origins. It is also easy to think of them as being predominantly an American initiative, but this not so.

The Mercury astronauts wore the first patches, but they were simply the insignia of NASA. Instead of mission specific patches, these pioneering astronauts gave their spacecraft names. The practice began when Alan Shepard named his spacecraft Freedom 7, the number 7 came not from the number of astronauts in the group as many have thought, but simply from the fact that this was the seventh spacecraft built. Subsequent crews named their craft with the seven suffix, and instead of the simple stencilled names on the spacecraft sides that Shepard and Grissom had, they came up with designs, logos if you like for their missions, with the help of an artist. These designs were much later made into woven patches, but they never existed in that form at the time of the missions.

Once the first crew had been announced for the Gemini program, Mercury veteran Gus Grissom, who would command the flight of Gemini 3, naturally wanted to continue the tradition of naming his spacecraft. He came up with the name "Molly Brown" after the Broadway musical of the time "The Unsinkable Molly Brown", clearly this was a reference to his Mercury flight that had ended up sinking. NASA officials thought that this name was inappropriate, and had been privately thinking for a while that this whole naming thing was getting out of hand, so they banned Grissom from using this name and demanded that he come up with an alternative. When he revealed that he rather liked the sound of "Titanic", they banned the future naming of spacecraft forthwith. NASA officials thought that the whole thing had been put to bed, but the next crew for Gemini 4 also wanted to commemorate their flight in some way, they had intended to name the spacecraft "American Eagle", but the recent banning had put paid to that. Instead, they decided that they would wear U.S. flags on the shoulders of their spacesuits, and every U.S. crew since then has done the same.

Mission patches officially came into being with the flight of Gemini 5, the crew of Gordon Cooper and Peter Conrad had already done battle with NASA Headquarters about naming their craft, and when they were also turned down they came up with the idea of a personal mission patch. It reflected the idea of U.S. military personnel having individual unit patches, and since the astronauts considered each crew to be a unit it seemed appropriate for each mission to have a patch. Conrad's father-in-law came up with the idea of a covered Conestoga wagon as part of the design, the idea being that it reflected the early pioneering spirit, and Cooper and Conrad added the slogan "8 Days or Bust" since that was the intended duration of their flight. Unfortunately, Jim Webb the then NASA Administrator, did not share the crew's enthusiasm, in fact it's fair to say that he lost his sense of humour over the whole thing. Both crewmembers pointed out that it was perfect for morale for the whole team of people involved in the flight to be able to wear such a patch. Webb saw their point but insisted that the slogan be covered up until the flight had successfully flown for that long, only at the end of a successful eight-day flight could they reveal it. The mission patch was here to stay, but NASA Headquarters insisted that they approve the design of every patch before it was made public, a practice that continues today. The naming of spacecraft made a brief re-appearance during the Apollo program when there would be two separate spacecraft flying at the same time, which needed to be identified by radio. Again, NASA Headquarters had to approve these names in advance.

All subsequent mission patches have featured the names of the crew, and imagery appropriate to the nature and objectives of the flight. Only six patches have appeared that did not contain any names at all. Gemini 7 and 10, Apollo 11 and 13, and much more recently, ISS Expedition 14, 15, and 16. This is becoming a more common practice with ISS missions, as many now routinely include several changing crew members. Expedition 15 has six different versions with different crew names, and this situation needs to be avoided in the future.

The appearance of names on patches has caused some headaches in the past, and indeed continues to do so today. In the early days of the space shuttle program, some crews decided to add the name of the particular shuttle that they were going to fly on board. Of course, this was a problem if the mission scheduling changed, and they were assigned a different shuttle, the patches would have to be changed. This particular problem came to a head with the flight of 41-E/41-F, which was originally assigned to fly Discovery, it was then changed to Challenger and renamed 51-E, and in addition, a payload specialist was added to the crew, his name was added as a tab sewn onto the bottom of the design. Unfortunately, for the patch manufacturer who had just completed these changes, a seventh crewmember was added, so they cut off the existing tab and replaced it with a new one with two names. This was not the end of the nightmare, however, since 51-E was then canceled, and crews jumbled around, the original core crew of 51-E remained, now given the flight 51-D, but one of the payload specialist had changed, and so had the space shuttle, it was to be Discovery again. The good news was that the manufacturer used the original 41-F Discovery patch, with a new tab sewn to the bottom. Because of all of these changes, 51-D was the last flight for quite some time that included the name of the

41E-F patch

51E Baudry patch

51E Baudry and Garn patch

51D Walker and Garn patch

shuttle, and all payload specialists tended to have their names on separate sewn on tabs.

On one occasion, the first shuttle flight to rendezvous with Mir, STS-63, caused some patch problems when one crewmember, Janice Voss, got divorced part-way through the approval cycle, the official patch originally said Ford, but was changed to Voss before any were produced.

The Soviet Union had also adopted the tradition of producing patches, but in a slightly different way to their American counterparts. In the Soviet system, cosmonauts have their own personal call sign, which they generally maintain during their entire career, the call sign of the commander of the flight is adopted as the main call sign for the mission. Therefore, patches have tended to be of a personal nature rather than a mission specific one. This has changed over the years, particularly when there is some special significance to the mission, for instance all of the Interkosmos international flights had a mission patch usually including the flag of the nation involved. The first known use of a personal patch was that used by Valentina Tereshkova during her Vostok 6 mission in 1963, it consisted of a white dove, and the letters CCCP. In fact, this was the first use of a mission patch by anyone, the U.S.A. not officially introducing them until Gemini 5 in 1965. Over the years, Soviet crews have worn a number of standard patches, many of them produced by Zvezda, who are the manufacturers of the crew's spacesuits. The patches produced by Zvevda have displayed the company's logo and the Russian word for Salyut, or Mir, and now

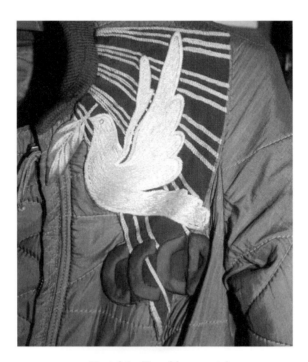

Vostok6—Tereshkova patch

ISS. Zvevda also produced the patch that was first worn by Alexei Leonov during his pioneering spacewalk, and later by the crew of Salyut 1. Since those early days, Soviet and Russian mission patches have been something of a mixture; many cosmonauts have carried their own personal patches, as well as patches that are specific to their mission. Many patch collectors have recently become dismayed at the sheer number of different patches that become available for just one mission. The more recent Soyuz taxi missions to the ISS have featured customized designs for each cosmonaut, often the same basic design, but with a different colour border for each crewmember.

Quite how patches will continue to evolve is unclear, Orion will carry crews of six at a time to the ISS, and the ISS standard crew complement is due to grow to six crewmembers at a time, it seems likely that mission specific or expedition patches may be on the decline, but astronaut/cosmonaut personal patches will increase. Time will tell.

Index

Printing: Mercedes-Druck, Berlin
Binding: Stein+Lehmann, Berlin